Conversion Concepts for Commercial Applications and Disposal Technologies of Energetic Systems

NATO ASI Series

Advanced Science Institutes Series

A Series presenting the results of activities sponsored by the NATO Science Committee, which aims at the dissemination of advanced scientific and technological knowledge, with a view to strengthening links between scientific communities.

The Series is published by an international board of publishers in conjunction with the NATO Scientific Affairs Division

A Life Sciences **B Physics**	Plenum Publishing Corporation London and New York
C Mathematical and Physical Sciences **D Behavioural and Social Sciences** **E Applied Sciences**	Kluwer Academic Publishers Dordrecht, Boston and London
F Computer and Systems Sciences **G Ecological Sciences** **H Cell Biology** **I Global Environmental Change**	Springer-Verlag Berlin, Heidelberg, New York, London, Paris and Tokyo

PARTNERSHIP SUB-SERIES

1. Disarmament Technologies	Kluwer Academic Publishers
2. Environment	Springer-Verlag / Kluwer Academic Publishers
3. High Technology	Kluwer Academic Publishers
4. Science and Technology Policy	Kluwer Academic Publishers
5. Computer Networking	Kluwer Academic Publishers

The Partnership Sub-Series incorporates activities undertaken in collaboration with NATO's Cooperation Partners, the countries of the CIS and Central and Eastern Europe, in Priority Areas of concern to those countries.

NATO-PCO-DATA BASE

The electronic index to the NATO ASI Series provides full bibliographical references (with keywords and/or abstracts) to more than 50000 contributions from international scientists published in all sections of the NATO ASI Series.
Access to the NATO-PCO-DATA BASE is possible in two ways:

– via online FILE 128 (NATO-PCO-DATA BASE) hosted by ESRIN, Via Galileo Galilei, I-00044 Frascati, Italy.

– via CD-ROM "NATO-PCO-DATA BASE" with user-friendly retrieval software in English, French and German (© WTV GmbH and DATAWARE Technologies Inc. 1989).

The CD-ROM can be ordered through any member of the Board of Publishers or through NATO-PCO, Overijse, Belgium.

Series 1: Disarmament Technologies – Vol. 14

Conversion Concepts for Commercial Applications and Disposal Technologies of Energetic Systems

edited by

Horst Krause

Fraunhofer Institut Chemische Technologie,
Pfinztal (Berghausen), Germany

Springer-Science+Business Media, B.V.

Proceedings of the NATO Advanced Research Workshop on
Conversion Concepts for Commercial Applications and
Disposal Technologies of Energetic Systems
Moscow, Russia
17–19 May 1994

A C.I.P. Catalogue record for this book is available from the Library of Congress.

ISBN 978-90-481-4880-6 ISBN 978-94-017-1175-3 (eBook)
DOI 10.1007/978-94-017-1175-3

Printed on acid-free paper

Table of Contents

Introduction

This book contains the lectures an contributed papers presented at the NATO Advanced Research Workshop on `Conversion Concepts for Commercial Application and Disposal Technologies of Energetic Systems`.
The meeting convened at Moscow, Russia, on 17th to 19th May 1994.

The workshop organized in collaboration with codirector Prof. B. Zhukov (Russian Acadamy of Sciences) was held

- to exchange information on the status quo efforts in the field of demilitarization of conventional surplus ammunition,
- to discuss the different approaches from Eastern and Western countries on the conversion of energetic condensed systems,
- to introduce new disposal techniques and reutilization routes which will solve specific conversion problems,
- to identify common research areas, where defined research tasks in cooperative partnerships can be initiated.

About 60 participants (30 % Russian) from 12 different countries took part (Czech Republic, France, Germany, Netherlands, Poland, Portugal, Russian Federation, Spain, Ukraine, United Kingdom, USA, and Usbekistan). The high international level demonstrates the relevance and importance of the workshop items. 28 oral presentations and about 30 posters gave an impressive overview of the conversion activities in the different countries. The conversion problems were discussed under the different stand points of scientists, industry and administration.

Additionally representatives of different Russian institutions observed the workshop (e. g. Department of Science and Technical Politics, State Committee of the Defence Industry, Defence Department, Trade-Industrial Committee and Municipality).

The opening remarks were given by Prof. B. Zhukov, Prof. H. Schubert (member of the NATO Disarmament Technologies Advisory Panel) and Prof. P. Sarkisov (Rector of the Mendeleyev University of Chemical Technology), followed by the introducing lecture of A. Rodionov (Deputy Minister of the Defence Industry).

The content of the given scientific papers can be grouped in three different conversion strategies:

- environmental friendly disposal techniques for energetic materials
- reuse and reutilization of energetic condensed systems from surplus ammunition
- civilian use of military laboratories and industrial facilities to develope new energetic systems for special commercial applications.

1. Disposal Techniques for Energetic Materials

In 1993 the use of sea dumping for disposal of industrial waste (including energetic materials) was banned under the Oslo Convention. Also the widely used practice of open detonation and open burning of explosives and propellants will be in the Western countries no longer accepted. The various Enviromental Protection Acts urge the governments to establish disposal facilities to destroy the surplus ammunition in an environmental friendly way.
Different incineration concepts are mainly proposed for such facilities. Installations based on fluidized bed incinerators as well as on rotary kilns are planned, are in construction or are already in use. Generally an after-burner and an expensive exhaust purification are necessary to reduce the air pollutant emission under the legal limits.

In 1991 the first incinerator fulfilling the legal regulations started to work in Germany. Two further installations will follow in 1994 (Rotter, FRG).

The Netherlands (van HAM) favor the fluidized bed incineration concept for the disposal of energetic materials. Also UK (Cumming), France (Tauzias) and Portugal (Fernandes) are planning similar disarmament facilities. Different water jet techniques are used for delaboration and preprocess of the slurry needed for fluidized bed incineration.

An alternative to incineration was introduced by A. Dahn (FRG) proposing the chemical/biological disposal of propellants. A simple chemical conversion delivering biodegradable products is followed by two biological process steps. The feasability was successfully demonstrated on a pilot plant.

Beside the disposal of conventional energetic materials there is a need to develope and to improve technologies to treat difficult 'items', e.g. pyrotechnic compositions containing hazardous materials. As most advanced technologies in this area the plasma arc pyrolysis and the super critical water oxidation (SCWO) were described and the risks and benefits of this methods discussed (Sotsky, USA). In USA and Germany SCWO pilot plants are planned or under construction to test the disposal of toxic/cancerogenic smokes, dyes and pyrotechnic compositions. Laboratory tests have already shown high destruction efficiencies on such hazardous materials (Krause, FRG).

2. Reuse and Reutilization of Energetic Condensed Systems

Facing environmental issues it, should be generally proved to recycle or to reuse the reclaimed energetic materials. Various contributions (USA, NL, P, CS, RF) dealt with the reuse of military explosives as commercial blasting agents (commercial mining, metal brazing, cutting). For these applications exits a limited market depending on the legal regulations in the different countries. A certain amount of TNT can be reused in civil mining explosives in spite of the fact, that using TNT for this purpose is a technical draw back (red water problem). Also PBX formulations for mining explosives and metal brazing are under development (Burch, USA). Powder shaped cordes were introduced to cut tanks withdrawn from service (Smirnov, RF).

The conversion of energetic condensed system into chemical raw materials is a further way to reuse energetic materials (Tartakovski, RF; Shilov, Ukraine). Examples are the chemical conversion of TNT to DAT, a raw material for plastic industry, the transformation of nitrocellulose to lacquer NC or the use of NC in combination with other polymer materials. During the last years, the disposal of the ammunition of the former East German army showed, that these concepts didn't succeed. For the market to such products is very limited and there are cheaper competing products which can be better produced regarding environmental aspects.

3. New Developments for Commercial Application of Energetic Condensed Systems

It is not easy to differ between real reuse of military explosives in commercial applications and new developments of explosives for special civil purposes. To my opinion the following examples presented by Russian scientists are more in the category of new developments.

So the production of artificial diamonds by detonative processes (Sakovich, RF) and the so called self propagating high temperature synthesis (Merzhanov, RF) of high melting compounds (nitride, carbide, silicide). These technologies demand very skilled scientists and a special development of explosives.

Another interesting application is the use of solid propellant magnetohydrodynamic (MHD) generators for mineral prospecting and earth quake prediction (Besruk, RF). The development of fire extinguishing systems based on gas generators is of great importance as alternative to the soon banned halone systems. Different systems based on propellants and pyrotechnic compositions were introduced as ready to sell products (Denisjuk, RF; Emeljanov, RF). But it was not clear, if the formed gases will meet the requirements for fire extinguishing systems in the Western countries.

4. Conclusions

The technical strategies in the Western and Eastern countries are completely different to solve the disarmament problems of surplus conventional ammunition. The Western countries are testing the utilization possibilities of the energetic components. In some cases military explosives can be reused as commercial explosives. Technical feasable concepts to convert energetic components to chemical raw materials failed in practice .

Environmental friendly disposal techniques are under developement and in part already in use to dipose of the energetic and hazardous components of surplus ammunition (incineration, chemical/biological, SCWO, plasma arc pyrolysis, etc.).

The Eastern countries, and especially Russia, are presently not very much involved in disposal technologies, which will reduce environmental risks. The research efforts are focused on new commercial applications for energetic condensed systems. It is not clear, how much energetic materials from surplus ammunition will be really used for the proposed applications. Some applications need very high developed energetic materials. Nevertheless, these concepts fulfill the conversion idea in reusing the former military facilities and the great know-how of the high skilled scientists.

The new applications for civil use led to several new cooperative partnerships dealing with e.g. fire extenguishing equipment risk assessment, biodegradation and ultra fine diamonds.

Dr. Horst H. Krause

Acknowledgments

I like to express my appreciation to Dr. Bajolan for his reliable assistance in organizing the NATO Advanced Research Workshop in Moscow.

Special thanks are due to all lecturers and authors for their cooperation in preparing their manuscripts, excellent presentation and dedication to the meeting.

Ms. Martina Kozka provided invaluable secretarial assistance in preparation for the meeting and Ms. Helga Dietrich and Ms. Karola Kneule for the expert typing of manuscripts.

Last, but not least, I like to express our gratitude to the Advisory Panel on Disarmament Technologies of the NATO Science Comittee for granting financial support for the meeting and Prof. H. Schubert for initiating and stimulating the workshop.

Energetic Condensed Systems

Directions of Conversion and Ways of Use for Determination of Actual Economic, Scientific and Technic Problems

A. A. Rodionov, I. N.Bocharov, B. P. Zhukov, A. V. Dranishnickov,
O. N. Ivanov

Russia.

Search and implementation of rational scientifically sound ways of conversion of defence industry are some of priority directions of state policy in Russia.

It is widely known that conversion is an expensive and long process big results of which could be reached only in middle-term or long-term perspectives.

The process of conversion was begun in our Country in 1989 and in scale and difficulty does not has any analogue in the worldwide practice. It is because of extremely high rate of decrease of volums of military manufacture (12...14 % per year contrary to normal standard 3...5 %) that resulted in decrease of share of military production in total volume of production of defence enterprises from 62 % in 1988 to 22 % in 1993.

Development of conversion is complicated by parallel process of radical reorganization of all economy of the Country and by transformation of economical space of the previous USSR.

The legal basis of business activities of enterprises of Defence Industry was established by the law taking in force from 1st January 1993 "About Conversion of Defence Industry in Russian Federation".

The most important stage of conversion development was the working up of "Russian Federal Programme for Conversion of Defence Industry".

H. Krause (ed.),
Conversion Concepts for Commercial Applications and Disposal Technologies of Energetic Systems, 1–10.

The purpose of the Programme is maximum use of scientifical, technical and industrial potential released in the result of Defence Complex Conversion for determination of important social and economic problems of development of Russia.

Within the framework of a federal programme are in realization 14 goal programmes including such areas of economy as transportation, communication, energy, energy supply, medicine, agriculture, chemical - forestry complex, ecology. For realization of the purposes it is proposed to find out 6 trillion roubles and about 22 milliard US dollars.

It is also developed "The Federal Programme of Industrial Utilization of Armament and Military Technics". It is known that energetic condensed systems (ECS) - powders, solid rocket, propellants, explosives, pyrotechnical compositions and the articles on their basis about which we will speak at our symposium are integral parts of all types of armaments from the simpliest ammunition till complicated rocket complexes and space units.

Vital cycle of every armament (scientific research - design - manufacture-exploitation) provides one or another method of completion of existence.

For ECS in past it was usually explosion or different versions of combustion. The scale of the started process of disarmament, the volume of the nearest future works in disarmament as ordinary as strategic arms and increased requirements of the protection of environment stipulate impossibility of ordinary methods of liquidation of arms.

An economical aspect of this problem is rather important. A state that invested considerable resources into development of arms and military technic preferes to receive with an assistance of scientists and engineers may be a partial compensation of realized expenses.

We can see two ways of determination of this complicated problem. One of them is development of industrial technologies of utilization that would satisfy contemporary regulations and requirements. Another one is use of unique properties of energetic condensed systems for determination of actual economical scientifical and technical problems in different areas of human activity such as manufacture of valuable chemical substancies, search and output of minerals, fire fighting, prediction of earthquakes, space opening up.

Not underestimating the importance of the first way we would like to emphasize considerable reserves and potential opportunities of the second way because of:

- nature of ECS as powerful autonomous sources of concentrated chemical energy with widely regulated characteristics;
- presence of economical goals that could be successfully determined using ECS;
- industrial potential of enterprises that manufacture ECS and the need to organize new jobs;
- high scientific and technical potential of defence industries of Russia.

Time limits of this report permit us to consider only the main directions and some results of the organised works.

In details many of the directions will be developed in key and stand reports, video films, samples of the exhibition, which you can see visiting Russian Mendeleevs Chemical - Technological University.

Explosives and Explosive Compositions

Explosives (E) as it is known are typical samples of energetic condensed systems of double purpose: big ammounts of the most widespread high explosive Trinitrotoluol manufactured during peace-time is used as main component of a lot of industrial explosives. Not accidentally manufactured under utilization of ammunition fusible explosive compositions (EC) on the Trinitrotoluol basis (TNT - aluminium, TNT-RDX, TNT-RDX-aluminium) are used in mining industries as waterproof granulated industrial explosives.

There is in a stage of development a technology of manufacture of catridgeous charges of unfusible EC on RDX basis (phlegmatized RDX - aluminium).

There were organized with positive results testings of shaped charges of EC (TNT-RDX) that are necessary for metalurgical works for crushing of steel and cast-iron massives.

It is developed the design of hose charges for crushing of big and top-size tyres of dump trucks (weight of a tyre up to 3.5 tones) that would permit to organize there utilization and to determine important ecological and economical problem in mining industry.

Future research and developments have the purpose to use EC extracted from ammunition in shaped cord-type charges for cutting of-metals, in cords for fire extinguishing ribbons in forest fire fighting, in special charges for reinforcement and welding of metals, in big stones crushing.

High sensitivity of industrial explosives manufactured on the basis of RDX-consisting EC extracted from ammunition is the factor that restrains their use.

Pyrotechnical Compositions

The use of pyrotechnical compositions and articles on their basis in civil purposes has wide range and rich history.

Well known such directions of works as:

- use of pyrotechnical compositions for holiday fireworks;
- use of pyrotechnical goods for hail-fighting and artificial initiation of precipitation.

Big practical significance has development of ecology safe pyrotechnical insecticidal and bactericidal means on the basis of thermosublimized aerosolproducing compositions. The means can be used for desinfection and fumigation of poultry farms, cattle-breeding and other industrial and domestic objects.

The generators of safety fire extinguishing substances are developed on the basis of pyrotechnical aerosol-producing compositions. They exclude the need to use halons and permit to develop systems of volume fire extinguishment that satisfy contemporary requirements.
The interdiction of halons manufacture from 1994 and future refuse of their use have stimulated the search of different ways of replacement of these ecology dangerous compositions in systems of volume fire extinguishment.

In one of the reports at our symposium there would be reported the results of the research on alternative version of determination of this problem and the products of combustion of special double-base type propellants are used like a working substance.

Powders and Solid Rocket Propellants (SRP)

It is known that energy of powders and SRP can be realised as in combustion as in detonation regimes in dependence of the character of the determined task and conditions of initiation.

Furthermore in some conditions the products of combustion of special SRP can become a source of powerful impulse of electric or light energy. Thus the nature of these type of ECP creates the prerequisites to there use for determination of a wide range of scientifical, technical und practical tasks.

Extracted from ammunition singelbase and double-base propellants nowadays have found the use as waterproof industrial explosives in a form of mixtures of granulated elements of both types without any change of their chemical composition (the energetic characteristics of these type of explosives are similar to characteristics of trinitrotoluene).

Futhermore the investigations showed that the exist technology of double-base propellants permits to manufacture industrial explosives with improved characteristics under some changes in there chemical composition.

Under the same technology can be manufactured universal elastic shaped charges of nitroglycerine powders, that are used in a lot of technologies (outting of metals, cutting of concrete monolites, explosion welding of metals of different type). The charges for signal rockets from special double-base propellants can be manufactured under the same technology.

The above-mentioned shaped charges are used for utilization of such metal-capacious and "not-technological" types of arms as ships, submarines and armoured technic. In this situation there is double utilization: there are utilized as arms as utilization body.

The perspective direction in ECS conversion is development and use in business of solid propellant gasgenerators and accumulators of pressure. There positive features are high reliability, compactness, constant readiness to exploitation, rapidity, autonomy.

We would list only some directions and areas of use of the technical samples that are in development:

- intensification of oil output because of thermo-gas-chemical treatment of near-well zone of a stratum by the products of solid propellants charge combustion;
- use of gasgenerators as sources of high-temperature compressed gas for tubes cutting in search and used wells;
- use of autonomous solid propellant sources of feeding in emergency systems of remote control of valves for gas mains;
- use of powder accumulators of pressure in fire extinguishing systems (local and mobile) for intensive displacement of fire extinguishing liquids from "Montague-type" tanks;
- use of powder generators of hot gas for rapid start of Diesel engines under minus temperatures (in Siberia, North, Arctica Region, Antarctic Continent);
- use of powder charges in emergency air-bags for cars.

The problem of transformation of chemical energy of solid propellants into electric and light one is of great theoretical and practical interest.

The development of special plasma propellants, the combustion products of which have high electric conductivity, permited to develop powerful impuls propellant sources of electric energy-magneto-hydrodynamic generators (MHD - generators).

The produced energy has found use in investigations of the Earth Crust on land and sea shelf, search of oil and gas fields and Earthquake prediction. The results of research and developments in this area will be reported at our symposium.

The search of ways of direct use of some arms that consist ECS for determination of not-military tasks is of great interest. It has been shown principal opportunity and high effectivity under some conditions of direct use of ammunition (artillery rounds, antimine charges) in mining industry.

The results of this developments are represented at one of the stand reports at the exhibition.

There is an experience of use of special propellant motors in building processes (production of vertical holes for foundations, horizontal channels for tubes and communications under railways, boring of wells in a field).

We estimate like a perspective the development of multipurpose rocket-space complex for taking into an orbit of small space systems with weight up to 1 t on the basis of disarmed military solid propellent rockets SS-20 and SS-25. There was information in the press about successful organization of the first start of this complex last year.

The developed in Russia technology of composite solid rocket propellents (CSRP) insures reliable manufacture of the charges of wide range in goals, sizes, shape with needed physical-mechanical and energetical characteristics for motors of military and commercial rockets, space systems, aerosol fire extinguishing means, MHD-generators. The equipment of this technology is used for commercial products manufacture: building materials (polymerconucrete), medicines, synthetic washing means, fertilizers.

Russian raw material basis permits to manufacture solid propellant charges and motors with ecology clear products of combustion on the basis of new chlorineless oxidizer - ammonium dinitronitrogen.

The problem of harmful influence of rocket starts on environment exists in the countries-developers and manufacturers of rocket technic.

In our opinion the starts in USA of Space-Shattle with two solid propellant booster motors 1000 t each are not so harmless. Every start is accompanied with output of 200 t of hydrogen chloride into atmosphere.

There is another big task in ECS conversion - the problem of development and realization of ecology safe methods of liquidation (utilization) of solid propellant motors with charges from wide range of military rockets. This task became very current after Russia and USA signing of International Agreement on Reduction in Stratagic Armaments (SNV-2).

It is known that liquidation of rockets of small and medium range (including SS-20) was organized mainly by explosion or combustion with rejection of considerable ammount of dangerous for human beings and nature components (hydrogen chloride, aluminium oxides, products of destruction of rocket motor cases).

Scale of work in future years and increased requirements on environment protection make us and as we think our partners find out and develop ecology safety industrial methods of utilization of Solid Propellant Rocket Motors.

This technically complicated an important problem needs in our opinion special discussions and we are ready to participate in the determination of the problem.

In conclusion we would like to attract your attention to the items of conversion of scientific research. In this area exist examples of interesting determinations.

Under the guidance of academician Vladimir Alexandrovich Tartackovsky, an outstanding expert in the area of ECS synthesis, there were found methods of processing of the most widespread high explosive-trinitrotoluene into polymers and dyes.

The scientists of one of siberian institutes under the guidance of academician Ghennady Victorovich Sackovich in co-operation with scientists of Institute of Hydrodynamics of Siberian Department of Academy of Science of Russia developed the method of manufacture of ultradisperse diamonds from carbon of detonating explosives.

Research in the area of combustion of explosives, powders and solid rocket propellants carried out during a lot of years in the Institute of Chemical Physics of Academy of Science of Russia under the guidance of Associate Member of the Academy Alexandr Grighorievich Merzhanov resulted in discovery of new higheffective method of production of refractory materials. Some other analogue examples exist.

In the present report there have been shown some ways and some results of conversion of ECS in Russia.

We suppose that starting today the International Symposium will enrich us with useful experience of the foreign colegues that work in the area of determination of the same tasks.

We hope to get more colleagues information on development of ecology safety technologies of utilization of ammunition, methods of decrease of sensitivity of explosives and compositions produced under these technologies, development of solid propellants on the basis of ammonium nitrate, commercial use of ammonium perchlorate and some other problems of mutual interest.

We consider this meeting as a step to co-operation of scientists of leading countries of the World for the purposes of successful determination of very complicated problems of conversion.

I wish all the participants and the guests of the Symposium good health and success.

DISPOSAL OF AMMUNITION IN THE FEDERAL REPUBLIC OF GERMANY – AN OVER VIEW

Gerald Rotter
Federal Office for Procurement and Defense Technology (BWB)
P.O. Box 7360
D-56057 Koblenz
Federal Republic of Germany

At the day of the reunification of the FRG an the GDR on October 3rd 1990, the few Federal Republic of Germany had two armies, the Western Bundeswehr and the Eastern Nationale Volksarmee (NVA), the National Peoples Army. The international treaties for the reunification limit the strength of the armed forces of reunited Germany to 370 000 men, much less than the western Bundeswehr with its 470 000 men had at that time. Because of that relationship of numbers there was no real necessity to introduce into service eastern weapon systems. That ment that the material of a well equipped whole army with a standing force of 160 000 men in 1989 war surplus.

Among this material were about 300 000 tons of ammunition, inherited from the former National Peoples Army.

Roughly 10 % of these 300 000 tons were of GDR origin, 90 % were imported from the Warsaw Pact States, especially from the Soviet Union.

Wereas the reduction of main armament systems takes place under the regulations of the Treaty on Conventional Armed Forces in Europe of November 1991, the disposal of the ammunition is purely a national task.

The reduction of this huge amount of ammunition had to start very quickly. The eastern ammunition depots did not meet western safety requirements and, even worse, the phasing out of the combat units of the NVA and the returning of their ammunition to the depots led to unacceptable storage conditions, at least in western eyes. In addition very often housing areas of the NVA were very close to the barracks, that means also to these units ammunition and were now within the safety range according to western fefinitions.

H. Krause (ed.),
Conversion Concepts for Commercial Applications and Disposal Technologies of Energetic Systems, 11–15.

The capacitiy for disposing ammunition in the old FRG was very low. Because of a planned consumption of aging ammunition, even life ammunition, there was only the necessity to dispose some 2000 tons per year, mostly done in a single installation.

In this situation it was quite natural to use the capacities of the former ammunition producing industry of the territory of the former GDR, that means in the five new Federal States. Besides that using these capacities close to the depots avoided the large scale transport of huge amounts of ammunition through a very densily populated country.

The GDR had produced small caliber ammunition, pyrotechnics, mines, handgrenades, a mosern type anti-tank missile under license from the Soviet Union, had had installations for maintaining rockets and for the reconditioning of large caliver ammunition. So there were the technical installations and also people with a thorough knowledge of ammunition and the immanent dangers.

These producers were:

- VEB Spezialtechnik Dresden with factories in Lübben (Spreewerke Lübben, small caliber ammunition), Königswartha (Mechanische Werkstätten Königswartha, small caliber ammunition, anti tank weapon RPG 18) and Pinnow (maintenance of large rockets, fabrication of the anti tank missile „Konkurs“;

-VEB Sprengstoffwerk Kapen with factories in Schönebeck and Gnaschwitz. This company produced several types of ammunition used by the engineers i.e. mines, handgrenades, charges etc.

In addition the NVA had a installation for reconditioning especially large caliber ammunition for tanks and artillery systems. In this installation old ammunition was taken apart and some components replaced which are limited in their lifetime i.e. fuzes, primers and propellant.

Nowadays the disposal of ammunition takes place on eleven sites. The capacity, depending of course from the type of ammunition, is in average 3000 tons per month alltogether. The overall situation is somewhat relieved because a considerable part of the ammunition is delivered to some other countries under military aid, so that the time frame will be end of 1995.

To use all these installations has had of course an additional social effect. It has given these companies the opportunity for restructuring and reorientating to the civilian market. At least some of them have done that with considerable success.

In addition to this existing sites some ammunition producing companies of the old FRG, losing their usual business because of the political changes, opened up new sites i.e. Rheinmetall in Weichensdorf and DASA (former MBB) in Steinbach/Neusorge. But nevertheless it turned out that the building of new installations is difficult. There are a lot of problems with the necessary approvals and of course with the people, very often green protesters.

All the former eastern companies are today owned either by West-German or American companies. Alltogether about 1300 people are employed by the main contractors for disposal of ammunition.

Ammunition contains in average 20 % of explosives and propellant. In the beginning the open burning of propellants was quite normal (as it is done all over the world), using official approvals of the former GDR authorities which under the treaty for the reunification are still valid. Of course one has to regard also the safety aspect: propellants outside the ammunition is more dangereous in terms of stability than inside the ammunition.

But in late 1991 the first incinerator, which meets the normal FRG standards went into service. Two other installations will go into service in the second quarter 1994. In addition there is an approved burning place for pyrotechnic components. These installations are completeley planned, it is well known that TNT can be recycled and either be used for military purposes again or be added to commercial explosives with some percent.

A considerable number of big charges, i.e. sea mines and different types of warheads, especially with some kind of mixed explosive were explded in the ordinary war in order to improve the safety in some storage sites very quickly.

One problem were the igniter caps of the small caliber ammunition produced in the GDR which contained mercury fulminate. Such type of igniter cap is not produced in western states for a long time. Because of lack of a rotary kiln incinerator fitted to handle the problem, it was decided to take apart the cartridges, literally by means of backward using the former producing machines and to recycle the mercury. So the environment may be grateful of being saved from several tons of pure metal mercury.

Another problem was the use of tetrachlorinecyclohexan (HCH), remainders of the LINDAN producing process. The tetrachlorinecyclohexan used in smoke tons which seems to be an old receipe, but nevertheless made enormous problems in handling under the regulations of the FRG especially of these tons which were damaged by sea water during use by the east-german navy. On the other hand the thermal dispoal offered no real problem. Sometimes it turns out that the avialable technical documentation and the hardware are not identical. The means one has be very careful during the disposal process.

Some studies, financed by the Federal Ministry for Defense (FMOD) were undertaken in the field of finding some alternative methods for the disposal of explosives and propellants. The results show, that the degradation by chemical and biological methods is possible, but also that the termal disposal is the safe way considering the short time frame. So, none of the projects has led to a full scale development.

The studies were

- Conversion of TNT to TAT (2,4,6 triaminotoluene). Contractor was the Chemie AG Bitterfeld-Wolfen in a cooperation with Rheinmetall. The study showed that it is practicable to converse TNT to TAT by means of a catalytic hydragenation process by using gaseous hydrogen and a noble metal catalyst. TAT can be used in a chemical industry for production to handle mixed explosive by separating the TNT from the other ingredients;
- Conversion of single base propellants. Contractor was Eilenburger Chemie-Werke GmbH. The aim of the study was to regain cellulosenitrate and reuse that CN in the chemical industry i.e. for CN-varnish. It turned out that problems are on the economic side because at least in western countries the time for that type of varnish seems to be over.
-..Conversion of single and double base propellant. Contractor was Berlin Consult (BC) in cooperation with Spreewerke Lübben and the Fraunhofer Institute ICT. This concept consists in a combined chemical-biological treatment of explosives in several steps. First decomposition of the propellant in hydrolysis under alkaline conditions. After that the explosive character of the substance is lost. The further biological decomposition of these hydrolyzates results in harmless gases and on a excess sludge which could be used for recultivating soils.

In the beginning, that means in 1991, money was not the most important matter. More important was to get things started. But in 1992, when all the installations were running, a competetitive bidding procedure was started. That is the normal way to deal with prices in a market orientated economy. And very quickly it turned out that prices could be cut to roughly one third.

Most of the costs are wages. Because of the very high wage level in Germany it is impossible to finance the disposal by selling the scrap, i.e. steel from shell bodies or brass from cartridges and other components. In Germany these net profits cover only 2-3 % of the costs.

The management for organising the industrial disposal of the ammunition is led by Federal Office for Procurement and Defense Technology, an oranisation whose normal task is to develop and procure material needed by the German Armend Forces.

DEMILITARIZATION RESEARCH AND DEVELOPMENT TECHNOLOGY FOR CONVENTIONAL MUNITIONS

Larry Sotsky
U.S. Army Armament Research, Development and Engineering Center (ARDEC)
ATTN: SMCAR-AES-P
Picatinny Arsenal, NJ 07806-5000

ABSTRACT

A conventional ammunition demilitarization technology research and development program was initiated at ARDEC in 1992 for the development of new and/or improved methods and technologies to treat "difficult" items and to maximize recycle and reuse of energetic materials. Three of the tasks currently being executed are application of the Plasma Arc Furnace, Supercritical Water Oxidation (SCWO) Technology, and Recycle/Reuse of Red Phosphorus (RP).

A state-of-the-art plasma arc furnace at a Department of Energy (DOE) facility in Butte, Montana is being evaluated to completely destroy and vitrify small munitions containing toxic/carcinogenic colored smokes, dyes, and pyrotechnic compositions in compliance with Environmental Protection Agency (EPA) regulations. Testing has been completed for a variety of pyrotechnic ordnance. Initial test results indicate that completely assembled small caliber and hand-held pyrotechnic items can be safely demilitarized in a plasma arc furnace.

SCWO is a relatively new technology being developed by numerous laboratories and private industry for flameless destruction of hazardous materials. This task is being conducted at Sandia National Laboratories (SNL) in Livermore, California. Initial testing on several colored smoke/dye compositions using SCWO technology has resulted in high destruction removal efficiencies (DRE). A contract has been awarded to Foster Wheeler Corporation to design, construct, debug/proveout, and operate a 100 gallon per hour production prototype unit.

Finally, an effort is being conducted at Pine Bluff Arsenal, Pine Bluff, Arkansas to evaluate the recycle/reuse of the red phosphorus/butyl rubber composition from the United Kingdom L8A1 smoke grenade to the newer version L8A3 grenade. A process for downloading smoke grenades and recycling of RP has been established. Qualification testing (smoke generation, accelerated storage) has been successfully

H. Krause (ed.),
Conversion Concepts for Commercial Applications and Disposal Technologies of Energetic Systems, 17–24.

completed. An Engineering Change Proposal (ECP) has been approved for implementation of the process.

1. Background

There is a large and growing inventory of conventional ammunition that requires demilitarization. The current stockpile is approximately 344,000 short tons (S/T) and could expand significantly due to the continuing downsizing and redeployment, base closures, and environmental restrictions. The inventory consists of a great variety of munitions items that are unserviceable, excess, and/or obsolete.

The method currently used for the bulk of the disposal activity is open burning/open detonation (OB/OD). This is the lowest cost, most widely available method at the various depots and ammunition plants having a demilitarization mission. Increasing restrictions on OB/OD due to Federal, State, and Local laws and regulations, as well as increased emphasis on pollution prevention, require that alternative methods be phased in as soon as practicable. The United States Army has developed a Demilitarization Master Plan which provides a road map in making the transition from primary reliance on OB/OD to resource recovery and recycling, with controlled incineration used for the balance. It is anticipated that OB/OD will be severely restricted by 1995 and will essentially be eliminated in the 1997/1998 time frame. This involves significant challenges and requires a variety of strategies and technologies to be pursued.

2. Plasma Arc Pyrolysis

Existing furnaces at Army facilities have been found to be unsatisfactory for demilitarization of small munitions items containing colored smokes, dyes and pyrotechnics. This is due to the fact that smoke and dye combustion products "lock-up" the fluidized bed incinerator and plug the pollution abatement equipment. Pyrotechnics generate high temperatures during combustion and damage the interior structural integrity of the combustion chamber.

The use of a plasma arc furnace is being evaluated to determine whether these problems can be overcome using this technology. The basic element is a plasma torch which utilizes a transferred electric arc process to generate extremely high temperatures which destroys organic wastes and transforms inorganic wastes into a glassy, non-leachable slag. It offers the potential for destruction and vitrification of toxic/carcinogenic colored smokes, dyes and/or pyrotechnics without disassembly. It is also being evaluated on other programs for treatment of contaminated soils and ash as well as for destruction of thermal batteries and other ordnance components.

DEMILITARIZATION RESEARCH AND DEVELOPMENT TECHNOLOGY FOR CONVENTIONAL MUNITIONS

Larry Sotsky
U.S. Army Armament Research, Development and Engineering Center (ARDEC)
ATTN: SMCAR-AES-P
Picatinny Arsenal, NJ 07806-5000

ABSTRACT

A conventional ammunition demilitarization technology research and development program was initiated at ARDEC in 1992 for the development of new and/or improved methods and technologies to treat "difficult" items and to maximize recycle and reuse of energetic materials. Three of the tasks currently being executed are application of the Plasma Arc Furnace, Supercritical Water Oxidation (SCWO) Technology, and Recycle/Reuse of Red Phosphorus (RP).

A state-of-the-art plasma arc furnace at a Department of Energy (DOE) facility in Butte, Montana is being evaluated to completely destroy and vitrify small munitions containing toxic/carcinogenic colored smokes, dyes, and pyrotechnic compositions in compliance with Environmental Protection Agency (EPA) regulations. Testing has been completed for a variety of pyrotechnic ordnance. Initial test results indicate that completely assembled small caliber and hand-held pyrotechnic items can be safely demilitarized in a plasma arc furnace.

SCWO is a relatively new technology being developed by numerous laboratories and private industry for flameless destruction of hazardous materials. This task is being conducted at Sandia National Laboratories (SNL) in Livermore, California. Initial testing on several colored smoke/dye compositions using SCWO technology has resulted in high destruction removal efficiencies (DRE). A contract has been awarded to Foster Wheeler Corporation to design, construct, debug/proveout, and operate a 100 gallon per hour production prototype unit.

Finally, an effort is being conducted at Pine Bluff Arsenal, Pine Bluff, Arkansas to evaluate the recycle/reuse of the red phosphorus/butyl rubber composition from the United Kingdom L8A1 smoke grenade to the newer version L8A3 grenade. A process for downloading smoke grenades and recycling of RP has been established. Qualification testing (smoke generation, accelerated storage) has been successfully

H. Krause (ed.),
Conversion Concepts for Commercial Applications and Disposal Technologies of Energetic Systems, 17–24.

Contaminated material is screw-fed into the rotating primary chamber and held away from the central discharge port by centrifugal force. It is heated by the molten slag, which is maintained at approximately 1,350 degrees Centigrade by the plasma torch until it is completely molten and oxidized. Additional oxygen is added to insure complete oxidation. Combustion gases exit from the primary chamber and enter the secondary combustion chamber where the temperature is controlled by a natural gas afterburner. As the gases leave the secondary chamber, they pass into a hot gas cyclone to remove the entrained particulates. The gases then enter the exhaust gas treatment system to quench the combustion gases, remove acid gases and entrained moisture and capture the remaining particulates. The scrubbed and filtered combustion gases are sampled and released to the atmosphere through the exhaust stack. A blower fan provides motive forces for the gases and keeps the primary chamber at a vacuum of 10 to 25 inches of water.

When the capacity of the primary chamber is reached, rotation is slowed and molten slag is gravity discharged through the water cooled throat and secondary combustion chamber into the slag collection chamber. There it solidifies into a glassy slag of low leachability. Typical system operating conditions are:

- o Primary chamber temperature: 1,100 - 1,400 degrees C
- o Secondary chamber temperature: 1,000 degrees C
- o Torch power: 400 - 500 kilowatts
- o Primary chamber residence time: 10 - 15 seconds
- o Secondary chamber residence time: 2 seconds (minimum).

The operational approach is to first reach the desired primary chamber temperature using a quantity of soil to serve as the heat source and diluent. The munitions items are then added individually into the feeder and dropped onto the molten soil at a controlled rate to maintain a slight vacuum and to keep the short duration pressure peaks below approximately 7 psig (the chamber has a relief vent set at 7 psig). At the completion of a run, the composition of the slag is expected to be about 50% soil residue.

The test data and analytical results from each run included analysis of the combustion products, particulates and leaching characteristics of the slag. Testing of a wide variety of munitions items was conducted during March-April 1993 in accordance with a test plan and an environmental permit. Typical items tested included the following:

o Signal, Ground, Smoke, Red, M168

o Grenade, Hand, Colored Smoke, Yellow, M18

o Signal, Illumination, Ground, White, M188

o Flare, Surface, Trip, M49A1.

Safety tests have been successfully completed. All end items were safely processed at feed rates tested. The pressure rise within the furnace (about 1 psi) was significantly less than anticipated. The glassy slag successfully passed the EPA TCLP test, indicating that the slag does not exhibit toxic characteristics. Particulate emissions were less than the Federal limit for a hazardous waste incinerator (<0.08 grains/dscf). NO_x emissions were also less than the Montana State limit (<7 lb/hr). However, several operational problems were encountered which require resolution prior to potential future implementation. These included inefficient collection of volatilized metals and particulate matter in the slag which resulted in excess filter maintenance. Finally, cost estimates and scenarios for potential future implementation are currently being developed.

3. Supercritical Water Oxidation (SCWO)

Large munitions containing organic-based carcinogenic smokes and dyes pose severe problems for demilitarization. OB/OD of these compositions produces toxic gases that contaminate the environment. The United States Surgeon General subsequently issued a moratorium (1980) prohibiting the OB/OD of munitions containing carcinogenic/toxic colored smoke and dye compositions.

SCWO technology (Figure 2) is being evaluated for destruction of carcinogenic/toxic colored smoke/dye compositions. Work is being conducted at the Department of Energy's Sandia National Laboratories (SNL) in Livermore, CA. The effort was initiated in 1992.

SCWO is a homogeneous phase, flameless oxidation process that occurs in water at temperatures and pressures above the critical point (>374°C and 3,200 psi). Organic materials which are relatively insoluble at standard temperatures and pressures are completely oxidized in the presence of water at supercritical conditions to provide high DRE's. Low operating temperatures when compared to an incinerator minimizes the formation of NO_x. The reactions are carried out in short stable tubular reactors where wastes cannot bypass the oxidation zone. The oxidation products are water (which can be recycled to the feed), carbon dioxide, and other small molecules which, dependent upon the feed, can be removed for potential recycle. The process basically consists of four operating stages:

- o Feed Pressurization
- o Pre-Heating
- o Mixing and Chemical Oxidation
- o Effluent Separation.

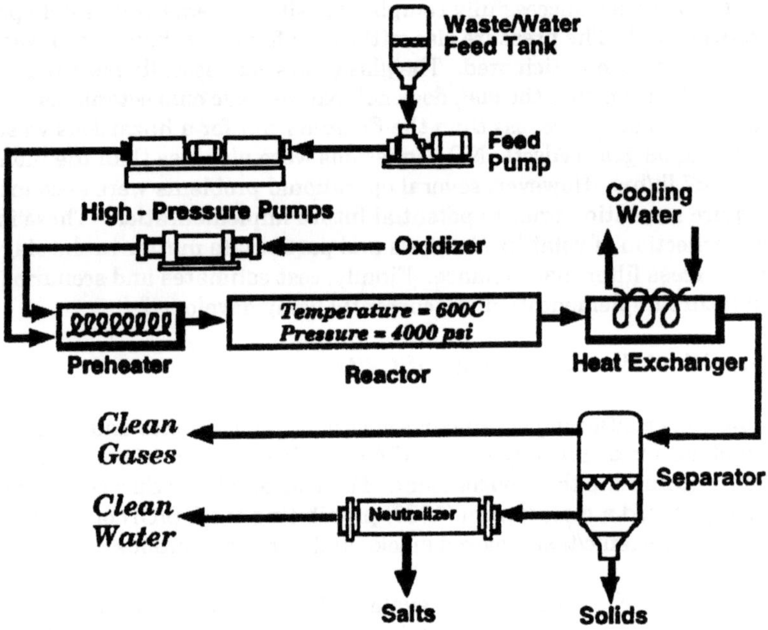

Figure 2

The waste material to be treated is mixed with water to produce a 5-10% aqueous solution or slurry. The slurry is pumped to the process pressure of approximately 3,200 psi and introduced into the pre-heat section, which is resistance heated. The oxidizer, which can be compressed air or oxygen, is also pumped in at this point, and the mixture is heated to 350-400°C, initiating the reaction. The temperature is increased to 500-600°C in the reactor section. The oxidation reactions occur within seconds after the critical point is reached. Most of the reactor residence time is required to increase the efficiency of destruction to 99.99+%. The hot effluent exiting the reactor is cooled in a heat exchanger and flows into a separation section to separate the solids, liquid, and gases.

The technical approach is to provide a SCWO system which can handle a variety of toxic and/or carcinogenic colored smoke/dye compositions. Technical challenges include the handling of potentially hazardous salts and selection of reaction system materials which will be reliable and safe (e.g., resistant to corrosion) in the reaction environment.

Typical smoke and dye compositions have been downloaded from stockpiled munitions at McAlester Army Ammunition Plant (McAlester, Oklahoma) and Pine Bluff Arsenal (Pine Bluff, Arkansas). Feasibility testing completed at SNL indicated very high DRE's (e.g., 99.69% based on Total Organic Carbon (TOC) analysis) at operating conditions of 560°C and 3,950 psi at a residence time of 10 seconds.

Development of a conceptual design package for a skid-mounted, transportable, 1,000 gallon/day unit has been initiated. The initial production prototype will be implemented at McAlester AAP (MCAAP). MCAAP is coordinating system permitting requirements with the State of Oklahoma Department of Environmental Protection. Implementation at MCAAP is planned for the late 1996 time frame.

4. Recycle of Red Phosphorus (RP)

The current demil inventory contains approximately 500,000 United Kingdom (UK) L8A1 smoke grenades (Figure 3) which are unserviceable. A program was initiated in 1992 to evaluate the recycle/reuse of the Red Phosphorus (RP)/Butyl Rubber smoke composition in a newer generation L8A3 smoke grenade.

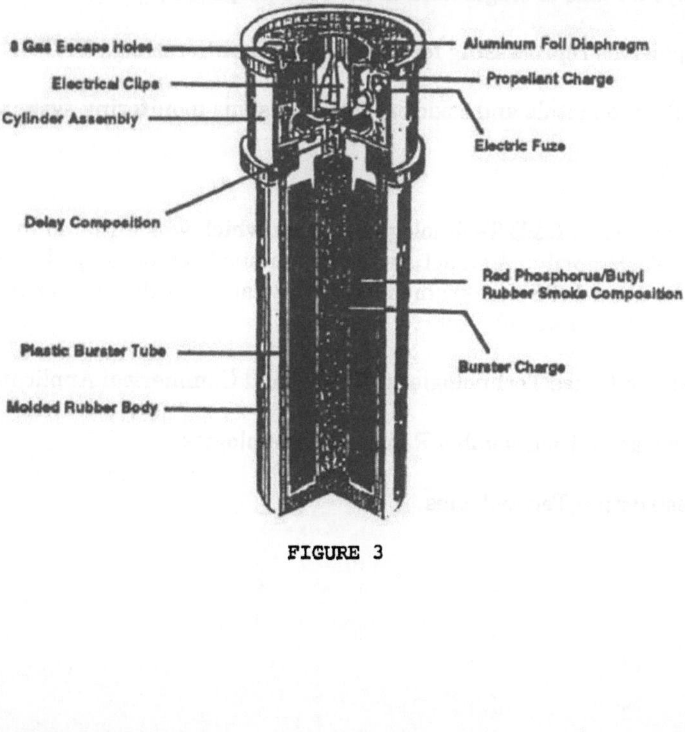

FIGURE 3

Work completed to date at Pine Bluff Arsenal has established and demonstrated a pilot plant process for downloading the RP/butyl rubber composition from the L8A1 and reprocessing the same for reuse in the L8A3. Accelerated storage testing (30, 60, and 90 days) at 160°F have been successfully completed. Smoke generation testing on grenades loaded with recycled RP showed comparable performance to standard L8A3 grenades. An Engineering Change Proposal has been approved to permit the reuse of the reclaimed RP/butyl rubber compositon in L8A3 smoke grenades. The process is currently being implemented at Pine Bluff Arsenal to meet an outstanding production buy for 160,000 L8A3's.

Recycle of RP from the entire inventory of L8A1's will result in a cost savings of approximately $4.9M. In addition, the Government does not currently have a dedicated supplier for RP. Recycle/reuse will prevent delays in meeting future production buys for the L8A3.

5. Proposed Future Efforts

The Demilitarization R&D Technology Program will continue to focus on developing new technologies for the recycle/reuse of munitions and materials to avoid disposal via OB/OD and/or incineration. Technologies proposed for future exploitation include:

- o recycle/reuse of recovered RDX and HMX-based explosives for military and/or commercial applications
- o recycle/reuse of single- and multi-base propellants
- o explosives reprocessing for cast loaded munitions (e.g., TNT)
- o real-time metals and toxic organic emissions monitoring system.

6. Summary

The Demilitarization R&D Technology Program which was initiated in 1992 is proceeding satisfactorily. A structured program has been developed and required efforts for the 1994-1998 time frame identified to address the needs and promising technologies in the areas of:

- o Recycle/Reuse Technologies (Military and Commerical Applications)
- o Separation/Disassembly/Removal Technologies
- o Destruction Technologies.

UTILIZATION METHODS FOR EXPLOSIVES WITHDRAWN FROM MILITARY STOCKS: DESIGNING, CARRYING OUT AND PRACTICAL IMPLEMENTATION

E. MILEWSKI, M.MISZCZAK, J.SZYMANOWSKI
Military Institute of Armament Technology
05-220 Zielonka, Wyszyńskiego 7, POLAND

ABSTRACT

The paper presents methods of disposal of ammunition that cannot be used for combat purposes. These methods allow for recovery both scraps of various metals and explosives. Utilization methods for small-arms, artillery and rocket ammunition as well as pyrotechnic components are discussed. Some usages of the above components in National Economy are presented.

1. Introduction

During a long-term storage, ammunition undergoes processes of natural ageing and "moral" wear. To identify in a right time an ammunition unsuitable for combat uses, its qualitative condition is monitored on a regular basis during its storage and basing on an optimized system of ammunition use in a peace time.

Diagnostic testing identifies lots of ammunition that have lost their suitability for use. These ammunition lots are transferred for destruction. To facilitate elimination of such an ammunition out of use, a system of mass disposal has been developed and implemented in Poland.

2. Utilization of small-arms ammunition

Because of a large build-up of stocks of small-arms ammunition withdrawn from use, economic methods of its utilization have been developed. Over a more than ten year long usage, the following disposal methods have been considered and tested:

- Mechanical dismounting method
- Thermal methods
- Disposal-by-firing method.

H. Krause (ed.),
Conversion Concepts for Commercial Applications and Disposal Technologies of Energetic Systems, 25–32.

2.1. MECHANICAL DISMOUNTING METHOD

This method is used for a larger calibre ammunition, i.e., 12.7 mm and 14.5 mm. Projectiles are separated from cartridge cases by means of special pulling machines or by breaking them out. Primer caps in the cases are fired by either mechanical or thermal method. Alternatively, they are pushed out using compressed air or pressure of water.

The above mentioned method uses manufacturing facilities withdrawn from ammunition production lines. About 25,000 rounds of ammunition could be dismounted this way within 8 hours.

2.2. THERMAL METHODS

2.2.1. *Disposal by means of electric arc*

This method uses initiation of flame-sensitive ammunition components by electric arc formed by a special installation. Initiation of primers by electric arc causes their separation from cartridge cases; quite often propellant charge does not get ignited. This is due to the fact that gas pressure generated by ignited primers forces the primers out before propellant charge ignition. Exposure of cartridge cases to a direct effect of electric arc results in ignition of propellant charge and total destruction of an ammunition round.

2.2.2. *Method using electric ovens*

This method is described by means of Fig.1. Heating oven (1) is made of steel metal sheet. It has ceramic internal surfaces, in which heating elements are fixed. Inside the oven there is a rotary drum (2) that has loading openings in its front part. Rear part of the drum has drainage holes (3) through which molten lead is discharged. Scrap metal is sent to a segregator (4). Disposal capacity of the installation is five hundred of 9 mm pistol rounds, one hundred of 7.62 mm rifle rounds or 80 igniters per minute.

The installation uses fume extraction systems (5) that include filters having replaceable filtering packs.Fume extractor output is 280 cu.m /hour.During disposal of 3,000 rounds of ammunition, approx. 35 cu. m of gases is generated (CO, H_2, CH_4, NH_3, nitrogen oxides). A dedicated installation for chemical absorption of NH_3 and nitrogen oxides has been developed.

2.3. DISPOSAL BY FIRING METHOD

This method consists in firing projectiles at bullet-catcher in form of a water tank. The firing is done by means of weapons withdrawn from use. This is both practical and economic method. Its disadvantage is a need to use separate facilities for lead melting-out and for disposal of special projectiles, e.g. of incendiary type.

Long years of experience have shown that the most economic and convenient from organisational point of view is the thermal method using electric ovens. This method, using a fully automatic small-arms ammunition disposal process, provides a complete safety for the personnel. It provides also a complete recovery of recyclable materials of ammunition components, i.e., cartridge cases, projectile jackets, lead, cores, etc. This method can also be used for disposal of small pyrotechnic components.

The thermal method is not used for disposal of small-arms ammunition of a calibre above 9 mm because of larger propellant charges and overall dimensions.

This ammunition is mechanically dismounted and individual pyrotechnic elements are destroyed separately.

Disposal of small-arms ammunition and pyrotechnic components is carried out in plants fitted with proper equipment that provides complete safety during operations of this kind. In such plants two types of disposal lines are used:

- Line for disposal of ammunition of calibre 5.6 - 9 mm
- Line for disposal of ammunition of calibre 12.7 - 14.5 mm.

The disposal process provides a recovery of scrap metal fully suitable for reprocessing in manufacturing plants.A multistage system of control of the recovered scrap metal has been introduced in ammunition disposal plants.

An example of advantages achieved by materials recovery at small-arms disposal is a case of disposal of 100,000 pcs. of 7.62 mm rifle rounds that gives approx. 1900 kg of brass and 1800 kg of steel and lead.

2.4. DISPOSAL OF SMALL PYROTECHNIC COMPONENTS (SQUIBS, DELAY ELEMENTS, IGNITERS, TRACERS)

Small pyrotechnic components as squibs, delay elements, igniters are destroyed in a facility shown in Fig. 2.

The principle of operation is use of effects of flames obtained by a burning fuel. Combustion chamber (3), hopper (1) and basket (2) are made as replaceable elements because of their fast wear. Housing (7) of the facility provides a space for expansion of gaseous products of decomposition. Feeder (5) is of a gravitation type. It is fitted with a protection against gas blow-out and effects of fragmentation. Fume extraction installation has filter (9) with replaceable pack. This installation provides a continuous process of disposal of small pyrotechnic components.

The above mentioned pyrotechnic components are also destroyed in an installation shown in Fig. 1.

3. Disposal of artillery and rocket ammunition

The research has been focused primarily on destruction and utilization of explosives, fuzes and propellant charges.

Following methods are used for explosive removal from projectiles: burning, melting-out, flushing-out or mechanical removal of explosives from the shells.

In the burning method, an explosive is ignited within the shell put on a properly prepared yard. Due to danger and detrimental environmental effects, this method is no longer used for mass disposal.

The flushing-out method is suitable only for removal of ammonium nitrate - based explosives; it uses water at a pressure of 9 - 10 atm.

The methods of explosive removal by melting it out of the shells are suitable only for removal of TNT and its mixtures. It should be emphasised that a large percentage of ammunition to be disposed off is filled with TNT or its mixtures with other explosives.

Disposal of a large-calibre ammunition is carried out in plant fitted with a proper equipment that guarantees complete safety. In such plants four types of disposal lines have been installed:

- Line for artillery ammunition disposal
- Line for disposal of rocket and mortar ammunition
- Explosive melting-out in dedicated installations
- Line for disposal of fuzes.

The disposal process produces a scrap (e.g. steel, copper, brass, etc.) that is fully suitable for use.

No method of mass utilization of fuzes has been devised to date. Disposal by dismounting is dangerous and its capacity is low. However, successful tests were run on use of removed booster charges to break oversized rocks in rock mining industry.

3.1. PROCESS OF MELTING OUT TNT AND ITS MIXTURES

A diagram of an installation for explosive melting out is shown in Fig.3.Projectiles are loaded into a melting chamber (1) and put on steam nozzles. The melting chamber is a steel structure; the heating medium is steam. During its operation the melting chamber is closed and sealed and all the elements being in contact with an explosive are heated. The chamber is fitted with a fume extractor duct. Molten-out explosive flows down and, via a discharge duct flows into a dosing channel (2) and next into a granulator (3) with a rotating water surface. The granulator is a metal container filled with water, in which an agitator rotates. Rotating water disintegrates a stream of liquid explosive into pellets that are conducted through sorter (4) to tank. Next the tank is transported out. Dimensions of the pellets are controlled by a thickness of the liquid explosive stream and speed of water rotation.

Water in the granulator is heated by molten explosive. To maintain water temperature at an optimum level for the granulation process, it flows in a closed circuit including a bank of sedimentation tanks made of concrete. The sedimentation tanks are cleaned on a regular basis. After a long time of operation, when cooling water is getting wasted, it is replaced by a fresh one by pumping water from the sedimentation tank to a storage reservoir and next to a special concrete tank. Deposit accumulated on the bottom is removed and burnt. The concrete structure protects effectively against water seepage into soil.

The melting chamber is fitted with a fume extraction duct with a filter to remove and adsorb fumes from inside the chamber. Also harmful fumes from the granulator are removed outside through filters having replaceable packs. Sewage system in the disposal installation premises is fitted with neutralizers of explosives.

3.2. TESTS AND USE OF THE EXPLOSIVE GRANULATE

The obtained explosives granulate features good physical and chemical properties. To improve some explosive properties of the granulate (e.g., to increase its sensitiveness to detonation effects), the granulate can undergo so called aluminizing, i.e., covering it with a permanent thin film of aluminium powder.

Results of some tests of TNT granulate are as follows:

- Bulk density - 0.949 g/cu.cm
- Melting point -77.2^{o}C
- Cook-off temperature - 265°C
- Critical diameter - 65 mm
- Capacity of a cavity blown in a lead block - 291 cu. cm

•Detonation velocity - 4400 m/s

•Chemical stability - comparable to industrial TNT.

Charges made of granulated TNT, with a dynamite detonator explode at a critical diameter of 40 mm and with TAT - 1 squib - at a critical diameter of 65 mm. Clean granulated TNT is less sensitive to initiation than the aluminised granulate or flaked TNT. In addition, the aluminium coating increases temperature and heat of explosion as well as sensitivity to detonation effects.

The obtained granulate features a low sensitivity to mechanical and thermal effects and to a flame. This makes it safe both in use and in transportation. Due to the above mentioned properties and proper detonation performances, explosive granulates obtained as a result of ammunition disposal have been widely used in rock mining. Despite its low bulk density, the granulate sinks in bores filled with water (its surface is porous and absorbs water) and easily detonates in such bores. Basing on a several-year long experience it has been found that, due to a medium brisance and relatively large working capacity, the TNT granulate can be successfully used in open pit mines. Quite often, to reduce costs of the explosives and reduce brisance, the granulate is used in mixtures with ammonium nitrate.

3.3. OTHER METHODS OF EXPLOSIVE UTILIZATION

In a case when uniform explosive charges can be removed from the shells, they are mechanically crushed. This refers to, e.g., explosive charges of mines and various types of explosive blocks. Crushing is carried out using dedicated crushing machines. The crushing process results in explosive grains of size from 1 mm up to 3 cm. Explosives recovered this way are used in rock mining.

When it is necessary to dismount an ammunition containing other explosives than TNT or its mixtures as an explosive charge, the charge is removed chemically.

3.4. UTILIZATION OF SINGLE AND DOUBLE BASE PROPELLANTS

The problem of propellant charge utilization was approached in a more detailed way in the beginning of the 70's. Until that time these propellants were destroyed by burning. In collaboration with rock mines, the problem of propellants' utilization has been solved by their use in open pit mines. After relevant Mining Authorities had granted a proper certificate, large quantities of propellants were sent to the mines. Various types of cutting machines were modified to cut propellants in form of long grains into smaller pieces. These machines provide both safety and high cutting capacity.

4. Conclusions

•Utilization of withdrawn ammunition is profitable either for economical or ecological reasons.

•Explosives, propellants and scrap of metal recovered during utilization can be used in National Economy.

•It is necessary to be developed mass utilization of fuzes and non-melting explosives.

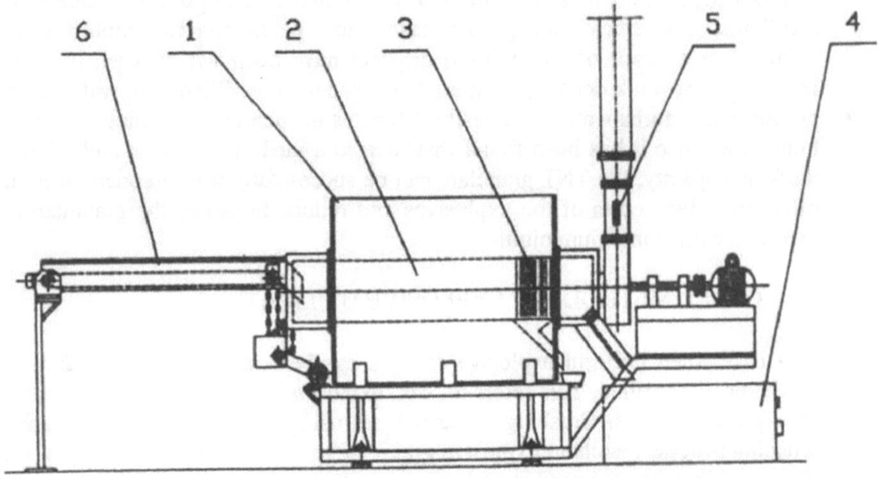

Figure 1. Instalation for utilization of small-arms ammunition.

1. Heating oven
2. Rotary drum
3. Drainage holes for molten lead
4. Segregator
5. Fume extraction system
6. Feeder

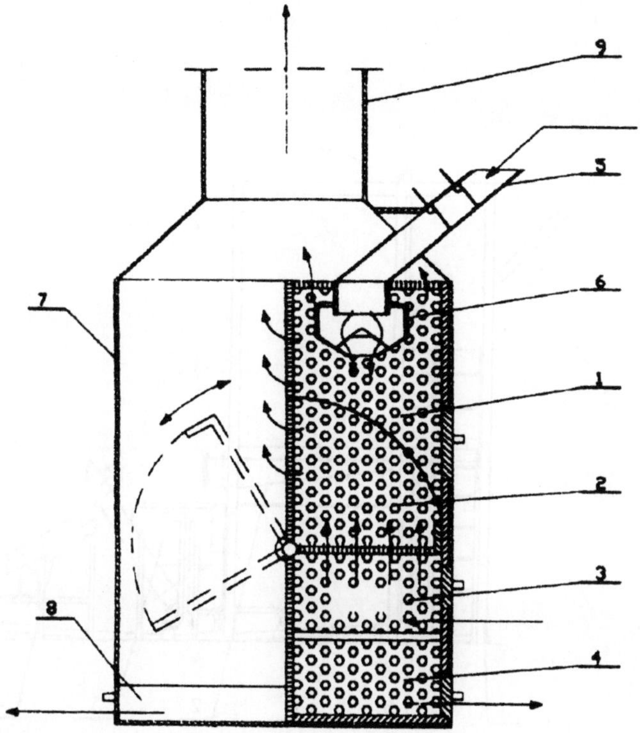

Figure 2. Installation for disposal of small pyrotechnics components.

1. Hopper
2. Basket
3. Combustion chamber
4. Ash pit
5. Feeder
6. Feeder shelter
7. Housing
8. Drawer
9. Filter

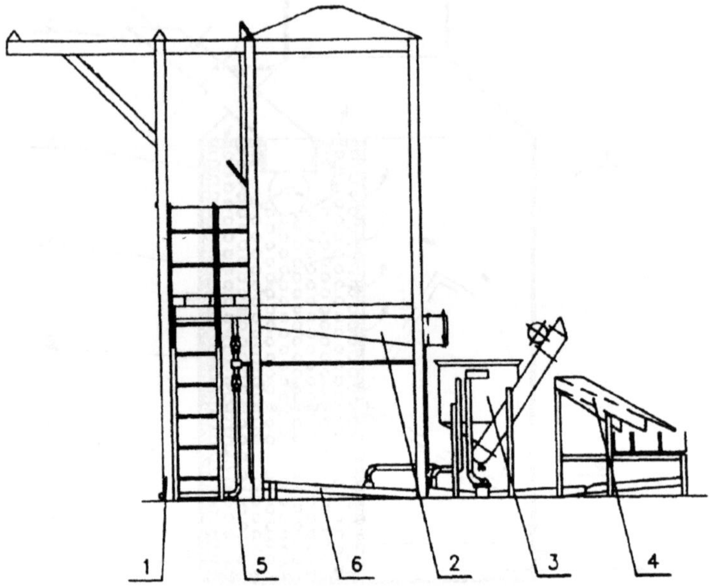

Figure 3. Installation for explosive melting out.

1. Melting chamber
2. Dosing channel
3. Granulator
4. Sorter
5. Steam system
6. Water installation

DEMILITARIZATION TECHNOLOGY AND EXAMPLES OF CONVERSION

Bernhard Brouqueyre, Jean Michel Tauzia

SNPE (France)

1. EVOLUTION OF SITUATION IN FRANCE AND THROUGHOUT THE WORLD

The destruction of outdated or overabundant munitions dates back to times of the earliest fire weapon industry.

In the past, and still today, most demilitarization operations are carried out with expeditive means (detonation, incineration, immersion in deep water).

Furthermore, the idea whereby munitions manufacturers are industrialists who not only manufacture and sell munitions but also maintain them operational and at the end of their life, handle their demilitarization is becoming more and more prevalent.

At the dawning of the 90's, the radical transformations that have affected the geo-political context, for France and many other countries, have led to an obligation to rethink and reorganize its defense policy and, at the same time, all the different nations have become aware of the need to protect the environment, henceforward implicit in the demilitarization process using clean technologies in which the recycling of materials with a high value-added factor is thoroughly incorporated.

In this new context, the destruction per se of ammunitions is only possible with a contribution from specific technologies taking economic and safety-related constraints into consideration.

2. EVOLUTION OF LEGISLATION PROTECTING ENVIRONMENT: THE BUILD-UP OF CLEAN TECHNOLOGIES

Awareness of the limitations and fragility of our planet has led public authorities to reinforce the legislation protecting our environment.

For instance, in the United States, the Resources Conversation and Recovery Act, put forward in 1989, encourages the recycling of raw materials whilst the Clean Air Act (1990) puts severe limits on wastes caused by human activity.

Similar limitations also exist in the European states, reinforced and completed by the Brussels directives which apply to all the members of the European Community.

In France, for the time being, dumping into air and water comes under the decree of March 1993 but industrialists are already being requested to think about new installations in order to meet the standards that the Brussels Commission is actively preparing for application as of 95-96.

In the following are a few examples of current evolutions demonstrating the toughening of legislation and which will set the yardstick for the technical and economic challenge that industrialists now have to face (the economic aspect is of particular importance).

Naturally, activities related to the implementing of energetic materials have to enter this line of thought.

H. Krause (ed.),
Conversion Concepts for Commercial Applications and Disposal Technologies of Energetic Systems, **33–39.**

Evolution of legislation concerning dumping into the air (example concerning energetic materials)		
	Applicable in France Decree March 93 Ministry of Industry and Environment	Applicable in France of Brussels Directive
Type of dumping into air	Limit concentration	Limit concentration
Heavy metals	0.2 mg/m3	0.05 mg/m3
Dusts (carbon)	50 mg/m3	5 mg/m3
NOx	500 mg/m3	50 mg/m3
HCl	5 mg/m3	5 mg/m3
Miscellaneous	--	0.1 mg/m3

3. THE CONDUCTING OF DEMILITARIZATION OPERATIONS IMPLEMENTING CLEAN TECHNOLOGIES: THE FRENCH RESPONSE

1. The founding of a specific industrial location

In the past, and today also, the destruction of obsolete munitions is carried out by modest-sized companies implementing reduced facilities and extensively using incineration in the open air with economic balance provided by the valorization of the recovered noble metals.

However, in some specific cases too (destruction of solid propellant motors), larger organizations (like SNPE), entered the field a long time ago.

Today, to cope with the needs for respect of the environment, two main players on the French scene working in the field of ammunition production and the implementation of energetic materials have agreed to join forces and industrial skills to form a new company, DEMUREX, the activity field of which covers the entire range of activities extending from dismantling to the recycling of ammunition of all types.

2. DEMUREX - .c. Implemented technologies

. DEMUREX: infrastructure

The chosen location is a 1400 hectare site in the south-east of France having a capacity to store 9000 tons of class 1.1 and 1.2 products according to French regulations.

Demurex is geographically placed in such a way that access by road and rail are easy. Demurex also enjoys nearby port facilities.

In addition, suitable security equipment is available on the site.

DEMUREX: the implemented technologies, the inherent means and the complement offered by the mother companies.

To cope with the extreme diversity of the objects to be processed, the list of which is too long to mention (shells, mines, rockets, smoke canisters, illuminating and incendiary munitions), and to process such munitions in a varying state of conservation (ammunition in good condition or oxidized), the site will gradually be equipped with appropriate means while ensuring that personnel safety is maintained at all times as is respect for the environment and the economical viability of the activity.

In the following we list the equipment in use or being produced, which will be completed gradually according to the expressed requirements:

- examination of ammunition (X-ray)
- cutting and deconfinement workshop using abrasive water jet
- dumping workshop using high pressure water jet
- fuze and igniter tube extractor shop, propellant dumping
- general purpose workshop designed for the disassembly of damaged ammunition or missiles
- incineration equipment for ammunition with a high probability of detonating (210 g TNT equivalent)
- incineration equipment for energetic systems and components of ammunition with a low probability of detonation
- treatment area for miscellaneous packages.

3. DEMUREX - Human skills

DEMUREX benefits from extensive design and research potential supplied by the mother companies.

Support is particular to two specific areas:

- safety studies applicable to the object, products and processes
- process surveys to evaluate the available technologies, select the most appropriate machines and dimension the industrial equipment.

In addition to direct support, the mother companies are carrying out more fundamental studies concerning, for instance, the transformation and recycling of energetic materials or research into new processes (e.g. biodegradation).

4. DEMUREX and the requirements of the French government

Essentially, the installations are dimensioned to cover national needs.

The needs are indicated, in general brush strokes, in the following table.

TYPE OF MUNITIONS	VOLUME TO BE PROCESSED BY WEIGHT (T)
Infantry ammunition	200
Small calibre artillery ammunition	100
Large calibre ammunition	2500
Mines	3000
Cargo ammunition	200
Gun Propellant	200
Rockets and rocket motors	100
Total	6300 tons

5. DEMUREX: it's position in the move towards demilitarization

By founding an entity that is sponsored by players well situated in the French munitions landscape, this project initially aims at covering French needs with clean technologies, but also of handling the tough economic challenge placed before it.

Indeed, much innovation, creativity and care will be needed to obtain a financial balance because if the point does not get due attention, the cost of dismantling might represent a significant percentage of the cost of founding.

In the perspective of research for optimum economy, DEMUREX is ready to collaborate with French or foreign companies in order to find with them, the essential synergies for bringing down the cost of the operations in question.

4. EXAMPLES OF THE RECYCLING OF ENERGETIC MATERIALS RESULTING FROM DEMILITARIZATION ACTIVITIES

The activities proposed below concern a variety of activity areas. Some have industrial applications, other simply involve operation at a pilot level but all are available for development at a larger scale.

1. Processing of rockets composite solid Propellant (industrial equipment currently being developed)

The destruction of solid propellant charges is traditionally carried out by igniting the motor in an area provided for the purpose.

This generates a large amount of hydrochloric acid, the dumping of which is today severely regulated; the process also generates noise which is detrimental to people living nearby.

To avoid these inconveniences, SNPE has put underway a major composite propellant processing program.

This process is based upon a sequence of individual operations (in terms of chemical engineering) as described below:

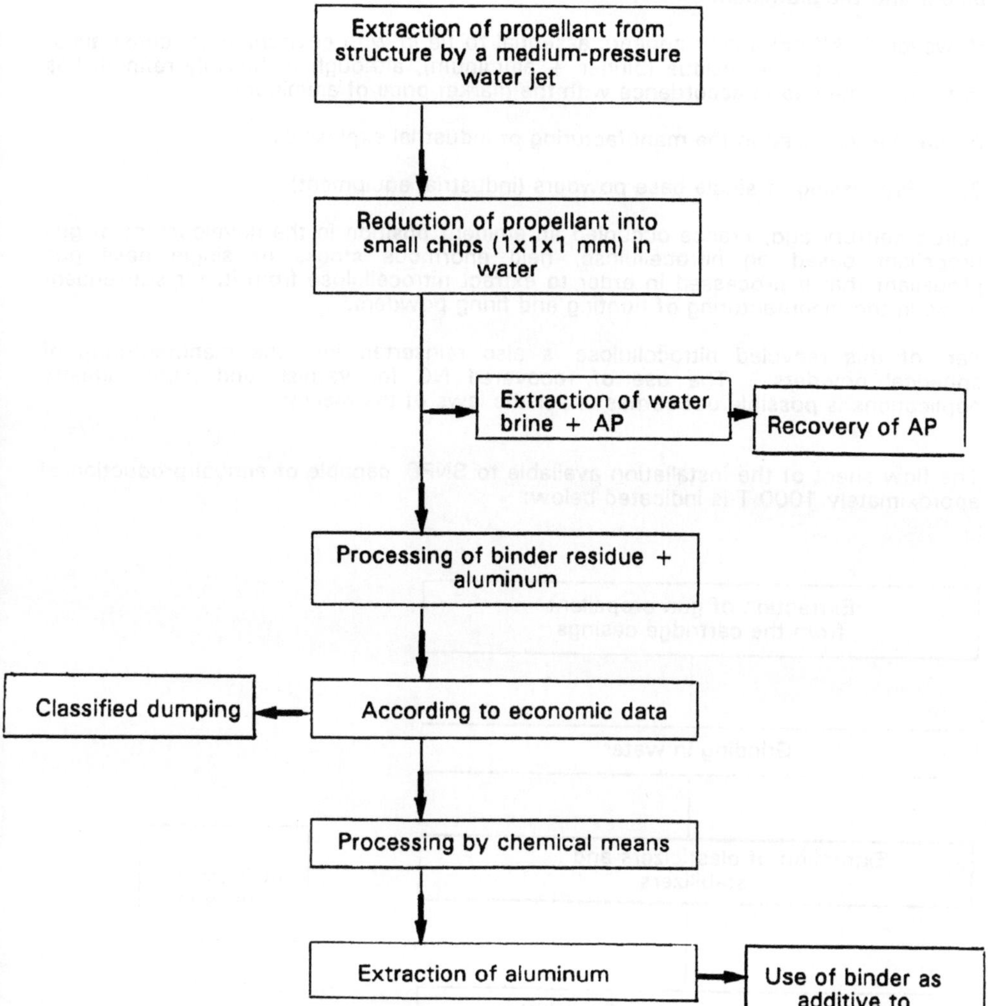

The proposed process, in general, is similar to that implemented in the United States by Aerojet and Morton Thiokol.

There are differences in the processing of the residue particularly in terms of the binder and the aluminum.

However, if AP extraction appears essential to tie in with environmental constraints, the treatment of the residue (binder + aluminum), although technically refined, has to be implemented in accordance with the market price of aluminum.

AP can be valorized in the manufacturing of industrial explosives.

2. Processing of single base powders (industrial equipment)

Just a century ago, France occupied an eminent position in the development of gun propellant based on nitrocellulose, held enormous stocks of single base gun propellant that it processed in order to extract nitrocellulose from it, for subsequent reuse in the manufacturing of hunting and firing powders.

Part of this recycled nitrocellulose is also reinserted into the manufacturing of spherical powders. The use of recovered NC for varnish and paint industry applications is possible but comes under the laws of the market.

The flow sheet of the installation available to SNPE, capable of annual production of approximately 1000 T is indicated below:

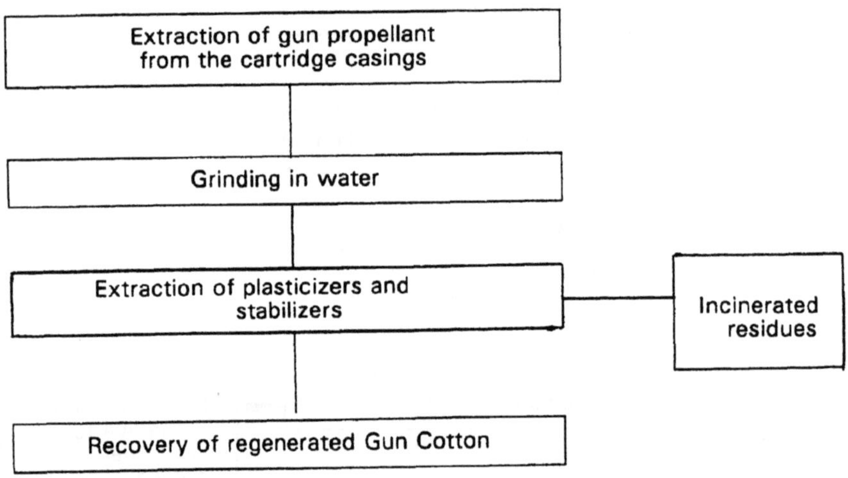

3. Processing for the recycling of liquid Ergols (IRFNA) - Pilot studies

Propulsion by liquid ergols uses smoking nitric acid and NO2 to which small amounts of hydrofluoric acid, iodine or phosphoric acid are added to inhibit corrosive of the containers.

The recovery of nitric acid, to be used for instance in the production of fertilizers (ammonium nitrate) presupposes the development of a process to extract inhibitor agents.

In cooperation with other organizations, a nitric acid purification process has been proposed. It uses, for instance, techniques developed for the purification of radioactive waste containing iodine that France has developed in the past for the nuclear industry.

The development of the process, defined at the pilot stage, represents no technical difficulties but the economic profitability, whose data will obviously be different from one country to another, if it ensures balanced operating treatment for the unit, will not cover the initial investment.

5. CONCLUSION,

The geo-political conditions of the time are moving towards a reduction in the volume of arsenals. Throughout the world, the destruction of surplus ammunition, of a wide variety of types, must abide by laws protecting the environment because public opinion is less and less inclined to tolerate noisy or polluting activity.

The technologies proposed to destroy ammunition today are numerous, offering a wide field of investigation open to researchers. However, only a few of them have reached maturity of sufficient development for industrial operation. Some of them are presented here.

By definition, the main baseline of this new activity, slowly disclosing itself to the profession, is that clean technologies are very costly and will not find a satisfactory economic balance through the simple valorizing of the recovered materials.

Accordingly, the financial contribution of public authorities is necessary if we want to preserve, in accordance with the obligations of the law, our environment. Demilitarization and recycling of obsolete ammunition are operations which must be financed by the owner states of the ammunition to be destroyed.

DEMILITARISATION AND DISPOSAL - UNITED KINGDOM STUDIES AND APPLICATIONS

A.S. CUMMING
Defence Research Agency,
Fort Halstead,
Sevenaoks, Kent, UK

Abstract

The UK has become increasingly involved in Demilitarisation and Decontamination in the last few years. Much of the UK need could be met by the traditional methods of burning and sea dumping until recently, but the increased appreciation of environmental hazards has led to the development of more modern methods. These are broadly similar to those employed in other nations but extend to the decommissioning and decontamination of former UK MOD sites.

The work carried out in the UK on both topics will be covered and progress reported, together with an indication of the likely direction of future developments.

1. Introduction

Demilitarisation has become one of the most important topics within the Energetics Industry. The reasons for this are not hard to find. The end of the cold war has left all the nations involved with stockpiles of weapons for disposal, and this when coupled with an increasing awareness of the need to protect the environment has given the scientific community a major problem.

It has become necessary to find new ways to dispose of the stockpile, methods that minimise cost and result in no damage to the fragile environment. This has become coupled with the need to close depots and other military sites while ensuring that proper decontamination has taken place. These are the immediate problems, though the awareness of them is leading to a requirement to include disposal as one of the criteria for proposing new energetic materials and systems.

The UK has not devoted as much effort to these studies as have other nations, particularly the USA and Europe. The stockpile was smaller, and traditionally remained in service for

H. Krause (ed.),
Conversion Concepts for Commercial Applications and Disposal Technologies of Energetic Systems, 41–46.

longer. The traditional methods of sea dumping or open incineration were until recently seen as satisfactory, and this view was reinforced by the practice of Crown Immunity, by which exemption from environmental controls was permitted.

However, this exemption no longer operates and the MOD has now declared its intention to adhere to the increasingly rigorous pollution regulations. Indeed these have bound UK industry for some time and thus they have developed capabilities only now being matched in MOD.

There are three aspects to be covered in this paper. The first is the decontamination and release of former MOD sites for other uses. The second is the development of incineration methods for munitions in UK, and the third is the development of the next generation of disposal/recycling methods.

It ought to be noted that the UK has developed considerable experience in explosives ordnance disposal during the last fifty years, experience that has continued to develop to meet the need to protect UK citizens from terrorist attack, and that this experience provides a background for future UK developments.

2. Decontamination and Site Clearance

In the last twenty years the UK MOD has closed many sites most of which had been used to store, manufacture or handle energetic materials. Some of these sites had been in use for centuries, in the case of Woolwich Arsenal since the sixteenth century. Such sites must now be certified as clear of explosives, which becomes harder to prove as the regulations become increasingly restrictive. In many cases operations have changed through the years and no accurate records are kept, perhaps because of wartime destruction. There is therefore an element of detective work involved in such operations together with an appreciation of the likely uses of such sites. Previous attempt to decontaminate must also be met and dealt with, in some cases resulting in the excavation of buried process buildings whose existence is deduced from old maps and reports.

Once usage has been identified as far as possible, and the site assessed, a clearance programme must be prepared. The nature and content of this will depend on the likely hazards but must include the taking of samples from buildings and soil for analysis to assess contamination; the use of ultrasonic and metal detectors for the location of buried items and the draining of water filled areas.

At Woolwich Arsenal, it was possible to burn out contaminated buildings to render them safe for demolition. However, although this took place only 10 years ago it would no longer be possible to use this solution, and less crude methods will be used at elsewhere, for example at Waltham Abbey, a research site now being decontaminated. When similar problems were encountered at a former Navy Depot, the internal building structure was removed for incineration.

The trace analytical techniques used have been fully developed and proven. At Woolwich soil samples were taken across the site at both surface and at depth through bore hole sampling. Sufficient information was gathered to permit the process to be completed to the level required at the time and to release the site for development.

The discovery of munitions during such a process must be dealt with on what amounts to a case by case basis. If the munitions are regarded as unsafe they are then dealt with on site using EOD methods and the residues taken for disposal. In the case of the Naval depot, charges had to be demolished using a donor charge which succeeded in initiating the uncovered rounds.

3. Development of Incineration Methods and their Use

On January 1 1993 the use of sea dumping for disposal of industrial waste (including propellants, explosives and munitions) was banned under the Oslo Convention. This was the primary method for the disposal of UK surplus ammunition. This included the disposal of EOD arising from area clearance. With the formulation of the EC Framework Directive on Waste (91/156/EEC) and the Environmental Protection Act (1990) it became desirable to review the UK position. It appears that neither provision prohibits the use of carefully controlled open burning/demolition. This control includes the minimising of pollution and smoke.

Items destroyed by open burning/demolition include those which cannot be safely moved, or the amount is too small to justify the cost of safe shipment or no contractors would wish to undertake the task. While disposal is carried out at authorised demolition grounds, ranges and MOD sites, the largest of which has no legal limits at present on the quantities so processed It is, however, assumed that this will not remain the case for much longer, and that other options be examined. A study was commissioned in 1989 to review the options, and this was reinforced by the formation of the Defence Environmental Working Group.

Thus despite a late entry into modern demilitarisation, the UK MOD is intent on developing a positive approach to the problem. Part of the approach consists of quantifying the UK problem, including assessing the number and type of munitions requiring disposal, the facilities already in place, including those provided by industry, and the assessment of the potential pollution problems. It was hoped that in this way it would be possible to prevent nugatory work.

The study indicated that while several MOD sites were involved in disposal, one particular facility stood out as the major disposal site, capable of handling high explosives, pyrotechnics and propellants. It has been decided that this facility will be developed by the building of an MOD Munitions Disposal Facility (MMDF) which will provide an alternative method for the disposal of those munitions being processed, though it is planned that it should remain complementary to industrial activities as UK policy is to dispose of ammunition through the use of a competitive tendering system. Several UK contractors have significant experience in the breakdown and reverse engineering of munitions and this would form part of the UK approach.

It is planned that the MMDF will begin operating in 1994 and consists of a rotary incinerator complete with an Integrated Pollution control. An Environmental Impact Statement is being produced as part of the licensing process and the facility will be gradually extended to permit more complex operations. DRA will be actively involved in the support of both operations and monitoring, including the assessment of the benefit to the environment of moving from traditional methods. This facility could thus be used to provide clean disposal for both UK and foreign materials.

As part of the planning of the improved facility a detailed study of the likely emission products was prepared. In part this was required to indicate by how far the permissible levels would be exceeded by such emissions. While such data exists elsewhere it was useful to try to quantify the UK problem. In addition the DRA at Fort Halstead have been tasked with providing support to the facility. This support includes emission monitoring, advice on safe reverse engineering and breakdown, as well as technical information about stores. As DRA has been involved with the design or approval for service use of most of the UK stores, it is logical to bring this experience into use.

As part of the exercise the number and types of store processed over the period of one year were measured and testing was initiated on real emissions. Recent tests on the during the disposal of large scale composite motors on the existing facility had not detected any unacceptable levels of HCl. However, it has been suggested that the next phase ought to include plume modelling and assessment in order to develop a better understanding of the risks and potential hazards.

While the UK may not appear to have the same scale of problem as is apparent in Eastern Europe or in the USA, our services operate a wide range of systems albeit in relatively small quantities. In terms of scale the UK ranks in Europe after Germany and Turkey and equivalent to France, Greece and Italy for its disposal needs. There will therefore continue to be a need for the development of a proper incineration policy. UK industry will be expected to share in the development of this and indeed has done to. ICI, which has considerable experience in dealing with environmental problems of this type has developed an international approach and is involved in the development of facilities in the US and elsewhere. Royal Ordnance is also involved, though mainly in providing the expertise in dismantling munitions amongst which are those whose design authority is RO.

Studies on more efficient uses of water jets have been carried out by DRA and the Royal Military College of Science. Studies have also been carried out into the uses of electrochemical methods for environmentally benign destruction, but this last will remain a future option.

It is clear therefore that the UK is developing incineration technology equivalent to that in place elsewhere. It is equally clear that it will be some time before the UK is in a position to match the facilities in use in the USA and Europe, but that there is a clear intention and requirement to do. What is less clear is the direction to be taken in future developments, and it is this that will be discussed in the last section of this paper.

4. Future Developments

The extension of the MMDF to provide a complete service both to UK MOD directly and through contractors, and to other non-UK customers, will bring the UK into line with present practice in other nations. However, the changing requirements for environmental protection have already initiated improved schemes elsewhere. There is, for example considerable work on the recycling of munitions, utilising recovered material for other purposes, and while this is simple for decontaminated metal or for easily broken down components it becomes more complex where applied to the body of the munition. One attraction of recycling is that it offers the opportunity to offset the cost of running a demilitarisation facility

There are two alternative approaches in development. The first, and simplest involves the recovery of the energetic ingredients and their use for other purposes with little or no modification. Such a process may be described as limited recycling. The material recovered after breakdown is not processed by used in some other activity, for example in mining or quarrying.

The second concerns the breaking down of the energetic material into its component parts and the reclamation of the original high value ingredients. The techniques for doing this are already available and use supercritical fluid technology with both water and ammonia for example. This could be the more valuable approach since the products have higher intrinsic value.

The first option is the easier. However its success depends on the size of the potential market. ICI, in a recent paper observed that the size of the civil market in the UK is relatively small, and is well served by low cost civil explosives. If demilitarised explosives and propellants are to be competitive in this market then the cost needs to match that of the existing satisfactory materials. If an export market is to be sought, then transportation costs will play a large part in the value equation.

The UK market has steadily moved from the used of sensitive materials to those sensitised by other ingredients. This would appear to limit the potential success of this option. However, it not one to dismiss lightly and will remain under review as a possible route forward.

The second option offers a route to reusable ingredients, apparently worth exploring for high value nitramines etc. However, these materials will need reassessment before they can be reused. The UK has stringent safety requirements for such ingredients administered through the Ordnance Board who would require to be satisfied about the safety and suitability for service of such materials.

Nevertheless, some work on the use of chemical technology for the extraction and purification of ingredients is planned, and the whole subject will be kept under review.

Finally, the need to consider future demilitarisation is being built into UK research programmes and formulation and processing studies are being initiated with the intention of emphasising the whole life of the energetic material, including service life and eventual disposal.

5. Conclusion

The UK is developing an approach to Demilitarisation which will match the present best practice elsewhere. The need to reduce surplus stocks in an environmentally responsible manner is a new requirement. Incineration remains the favoured approach at present though other future options are being reviewed and a process of continuous assessment developed. It remains UK policy that any programme will include significant industrial involvement, and that this will both benefit the UK and assist other nations who may not possess an environmentally responsible national facility.

6. References

Brown D W, ICI Explosives Environmental :
Demilitarisation in the UK - A View from Industry
ADPA Advance Planning Meeting for Industry, Arlington Virginia, USA, May 1993

Hindle M R, DEWS, MOD, UK :
Demilitarisation of Royal Naval Armament Depot, Bull Point
ADPA International Symposium on Energetic Materials Technology, Orlando Florida, USA March 1884

MUNITIONS DISPOSAL AS A TOTAL CONCEPT

DRS. N.H.A. VAN HAM
TNO Prins Maurits Laboratory, PO Box 45,
2280 AA RIJSWIJK, Netherlands

Abstract

Conventional methods for the disposal of munition and explosives like dumping and open pit combustion can no longer be tolerated in view of new environmental protection laws. For this reason TNO has investigated some alternative methods for the disposal of munitions. Starting with the reversed assembly of munitions, followed by the separation of explosives and metal parts, one can study possibilities for recycling these components.

From our investigations it turned out that controlled combustion is the most mature, promising and universally applicable technique. Controlled combustion makes use of a closed furnace system; most promising for the situation in the Netherlands seems the fluidized bed furnace. Additional scrubbing systems (dry chemical / wet) are needed to remove the remaining hazardous products like HCl, SO_2, NO_x.

Calculations for a complete disposal unit were made.

H. Krause (ed.),
***Conversion Concepts for Commercial Applications and Disposal Technologies of Energetic Systems*, 47–54.**

1. Problem description

The disposal of munition and explosives is in fact the last phase in the life time cycle of these ordnance items. To put them in the right perspective it is necessary to overview the whole cycle.

Procurement-phase:	• tactical study	
	• demands	
	• trials	
	• comparative bids	
	• contract	
In service-phase:	• transportation	
	• storage	
	• trials	
	• exercises	
	• usage	
	• maintenance	
	• modifications like:	• other fuze
		• other warhead
		• exercise from HE
End of life time-phase:	• sell	
	• reversed assembly	
	• recycle	
	• disposal	

For the most practical solution of the disposal problem it is necessary to overview the whole life time cycle of the munition. From the viewpoint of disposal it is strongly recommended that reverse assembly of the munition and subsequent recycling of the components is feasible. Preferably this aspect should be taken into account during the procurement phase. Of course this is not the most important munition issue; first of all the munition must have the optimum power to stop the enemy, second demand is maximum safety for own personal during handling transportation, storage and usage, third demand is long life time, fourth non toxicity; munition must not introduce a labor hygiene problem to military personnel, last demand will then be the possibility for recycling.

2. Pre-treatment

A good separation between the explosive and the metal parts of the munition is mandatory for a more economic disposal procedure. In this way materials can partly be reused and the controlled combustion will be more efficient.

Reverse assembly or disassembly is the conventional pre-treatment of munition: it includes the separation of the projectiles from the cartridges, the defuzing of the projectiles and the removal of the explosives from the projectiles. The reverse assembly is applicable to most regular munitions.

Water jet cutting is a powerful variance on the conventional metal cutting tools. By means of a high pressure water jet and some abrasive, munitions can be cut with high speed. Nevertheless it is a relatively safe method as the water cools the metal and possible ignition of the explosive is suppressed.
This technique can be utilized on those munitions that cannot be opened by reversed assembly (non regular munitions), e.g. munitions in bad condition or munitions that originate from World War II and have been buried in the ground for a long time.
The system consists of a water jet cutting system. The water with possible explosives is collected. After filtering the water can be used again. This technique has been developed by TNO and tested in practise by the Dutch Navy. TNO has tested the water jet cutting technique with all kinds of munitions such as 40 mm HE, 105 mm HE, 155 mm HE, 81 mm mortar with pyrotechnic flare composition (Mg!).
The deliberate cutting action on very sensitive parts like the fuze and the igniter did not result in a reaction, so the safety of the system is proven in practise.

Figure 1. Test set-up for water jet cutting of 155 mm projectile.

Figure 2. View of 155 mm projectile after water jet cutting.

3. Slurry preparation

The High Explosive fillings are washed out the munitions by means of a water jet technique. The explosives are milled under water till a particle size of 50-1000 m is reached. The propellants are treated in the same way, but the water/propellant slurry is stored separately.
The sensitive (mostly inorganic) explosives from the igniter and the fuze are also separated from the metal parts with the water jet and collected for chemical conversion.
Both the High Explosive slurry and the propellant slurry are transformed into the desired concentration of 10-50% explosive. This is realised by filtering techniques. In order to prevent the settling of explosives from the slurry, additives are used to create a stable slurry. Also a pumping system is used to prevent the explosive particles to settle.
The explosion safety of the installation is realised by the design of the pipes necessary to transport the slurry from the storage to the fluidized bed furnace. The diameter of these pipes must be under the critical diameter for a stable detonation of the slurry in the pipes.
Special attention is paid to the injection of the slurry into the fluidized bed furnace. The injection nozzle is dimensioned to create fuel droplets that are completely combusted before they can reach the upper part of the bed.
The storage of the high explosive / water-slurry is very safe as this slurry can only detonate if initiated by the powerful shock of a donor explosive.

4. Fluidized bed combustion

The principle of controlled combustion is the well defined feed of fuel (slurry from explosives and water) to a closed combustion chamber or furnace. In this way the explosives can react portion by portion with excess air to give the cleanest combustion products. This in connection with the further treatment of the reaction products to satisfy the threshold values defined in the National Environmental Protection Laws.
On the other hand the amount of explosives present in the furnace at the same time can be regulated, to avoid pressures that can damage the furnace.
A fluidized bed furnace (see figure 3) makes use of a flow of hot air trough a packed bed of silicon oxide particles (sand).

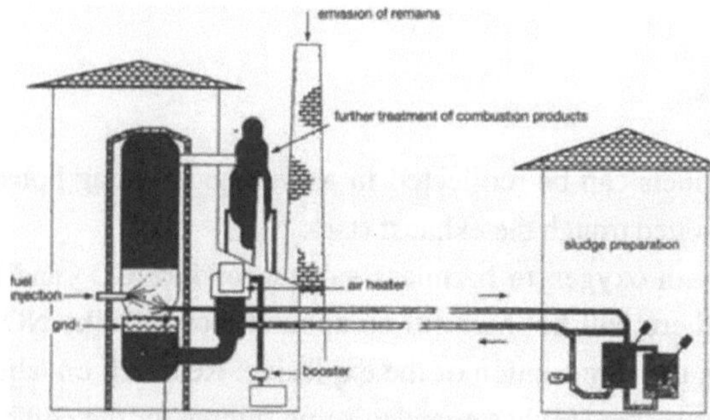

Figure 3. Principle of fluidized bed furnace.

Due to the action of the air flow the particles of the bed are floating and act like a liquid. The fuel is injected in or above this floating bed in the form of fine droplets ensuring optimum mixing with air. It is of importance that the particles have reacted completely before they reach the top of the bed. A CO-monitor is installed to control the proper functioning.
Catalyst like NiO can be added to the bed to facilitate the decomposition of explosives and to suppress the NO_X formation. The bed material can be varied in composition between pure silicon oxide and aluminium oxide. That is the reason that fluid bed incineration has often been used for the clean combustion of waste.
In order to be applicable for the disposal of munition pre treatment of the munition is mandatory. A combination with water wash out or water jet cutting systems seems attractive; the explosive is separated from the metal parts and grinned under water to a slurry. The slurry is collected in a reservoir, the slurry is stabilized by suitable agents

and constantly stirred to obtain a homogeneous slurry. The transportation system is in fact a loop between the mixing vessel and the furnace. Due to the efficient and complete combustion, the use of an afterburner can in most cases be omitted.
An advantage of the fluidized bed furnace over a rotary retort is the energy balance: due to the efficient combustion (especially when catalyst are used) and the possibility to use the excess heat from the bed for preheating the air and energy recovery. Another interesting possibility is the large heat capacity of the bed; when this has reached its working temperature (500-800 °C) there will be no need for extra heating other then generated by the combustion of the explosive in water slurry. When the process is stopped the bed remain at high temperature for a long time, even over the weekend. After 16 hours the temperature is still 600 °C so the furnace can be started up easily at low energy cost (thermal fly-wheel).

5. Exhaust cleaning

The solid reaction products can be collected in a cyclone or a bag house, while the gaseous products are emitted trough the exhaust stack.
Explosives can react with oxygen to harmless substances like CO_2 and H_2O for the mayor part. However there will be harmful compounds formed like NO_x, CO, HCl, HF, SO_2, depending on the composition of the explosive. Research on laboratory scale by ICT Germany (Dr Volk, see [6]) has revealed some figures for the combustion of the most common explosive materials.

TABLE 1. Harmful exhaust gases due to the combustion of explosive (w %)

Explosive	NO_x	HCl	SO_2	H_2S
Single Base	10			
Double Base	19			
TNT	0,01			
Rocket Prop.		25		
Black Powder	-	-	0,1	6-8

At TNO-PML we have found for TNT 3-7 weight % NO_x and for a Single Base Propellant 2-3 weight %; it is obvious that the reaction is strongly dependent on the conditions in the furnace.
In the Netherlands the gaseous exhaust products should fulfil the limits given in table 2.

TABLE 2. EC Exhaust Threshold Limits

Component	Threshold value (mg/m^3)
Total solid dust	5
HCl	10
CO	50
SO_x	40
NO_x	70
Heavy metals	1
Cd/Hg	0.05
PCD's (dioxines)	0,1 nanogr. TEQ/m^3

Removal of CO/C_xH_y. This can be realized by correct functioning of the furnace and its afterburner section: The temperature in the afterburner section should be well above 850 °C, the % oxygen should be above 6 vol. %, the residence time of the reaction products in the afterburner should be above 2 seconds.

Removal of NO_x. By careful design of the furnace burners and the use of clean fuels such as natural gas, the excess formation of NO_x due to the reaction between N_2 and O_2 can be suppressed. Furthermore, NO_x can be decomposed by the use of catalysts. This is most conveniently realized in the fluidized bed furnace. Other possibilities are the chemical binding of NO_x with NH_3 and the wet scrubbing technique.

Removal of HCl, SO_x. This can be realized by use of a wet scrubber in combination with a chemical scrubber (Na_2CO_3).

Removal of dioxines. During the combustion processes of explosives, all the necessary conditions are present for the formation of dioxines. Suitable techniques for the removal of dioxines are: injection of active carbon together with CaO, and the use of active carbon filters.

6. Cost estimate for complete design

In table 3 the following calculation was made for the situation in the Netherlands: all prices in Kfl. [7].

TABLE 3. Cost estimate for the construction of an explosives incineration plant for 100 kg/hour

Element	Fluidized bed	Rotary retort furnace
furnace	1000	4500
water jet cutting	220	220
mill for fireworks/ munition	100	
control app.	300	150
scrubber(wet)	175	175
denox(chem.)	50	50
dioxine filter	50	50
site preparation	1155	1355
total cost	3050	6500

The investment cost of the furnace dominates the total cost; the fixed capacity of the rotary retort exceeds the annual Dutch need many times, whereas the fluidized bed furnace can be tailored to the actual needs.
For the fluidized bed the annual exploitation costs are calculated at 1000 Kfl.

7. References

1 Kearney, P.C. et al., (1983) *Chemosphere* **12**, 1583
2 Hoffsommer, J.C. et al. Report NSWC/WOL-TR-77-146
3 Maby, W.R. et al. (1983) *Chemosphere* **12**, 3
4 Kaplan, D.L. et al. Natick TR-81/019
5 Hill, D.B. (1980) *Explosive Safety Seminar* **19**, vol II, p. 995
6 Volk, F. Chemistry and Physics of Energetic Materials, Kluwer 1990, p. 511
7 Bartelds, H., Temmink, H.M.G., TNO-IMET Report 91-453 *Haalbaarheidsstudie verbrandingsinstallatie voor het explosieven opruimingscommando* (in Dutch)

NEW TYPE OF ARTIFICIAL DIAMONDS AND PHYSICAL - CHEMICAL FUNDAMENTALS OF THEIR CREATION

G. V. Sakovich, E. A. Petrov, V. F. Komarov, N. V. Kozyrev
NPO "Altai", Biysk, Russia

An analysis of experimental results, carried out by NPO "Altai" jointly with the Institute of Hydrodynamics, Siberian Branch of Academy of Scienes, has given rise to the new technical and scientific field of ultrafine diamond synthesis (UFD) right from carbon of high explosives (HE). When doing experiments there has been registered a abnormally high output of diamond-like carbon phase, obtained at dynamic loading of organic carbon-containing substances. This fact made it possible to invistigate thoroughly all reason of the process and in 1982 there was finally created a new method of ultradispersed diamond synthesis from explosives. (1,2). In 1988 they learnt some information about the synthesis of ultradispersed diamonds from explosives, carried out both by US scientists and German ones.

Interest in the explosively obtained diamond particles is not accidental. Pressure of hundreds of thousand of degrees, specifying detonation, accommodate the field of thermodynamic stability of a diamond phase in a diagramm of possible carbon states (Fig. 1). But successful method realization became possible only when taking into account the importance of kinetic and thermodynamic processes of carbon diamond phase formation in the chemical reaction zone of detonation wave. Scientific technical trend of UFD production from explosives includes handling the following problems:

1. Proper synthesis of diamond particles in the chemical reactions zone of detonation wave.
2. Preservation of UFD during of secondary chemical reactions and physical processes, occuring in detonation products while expanding and cooling in the explosion chamber.
3. Trapping of supersmall particles (20 - 100 A) of condensed phase.
4. UFD isolation from condensed products of the detonation synthesis.
5. Investigation of UFD properties and structure.
6. Search for most effective fields of UFD application.

SYNTHESIS OF DIAMOND PARTICLES IN THE CHEMICAL REACTION ZONE OF DETONATION WAVE

Formation of diamond particles in a chemical reaction zone occurs at detonation decomposing of a large number of single-component and composite explosives and powders (Table 1) (Fig. 4,5).

H. Krause (ed.),
Conversion Concepts for Commercial Applications and Disposal Technologies of Energetic Systems, 55–72.

To synthesize diamonds in detonation wave from proper carbon of the explosives it is necessary to meet two conditions:

- detonation products of explosives are to contain condensed carbon (CC);
- pressure and temperature in detonation wave are to fit the field of thermodynamic stability of diamond phase of carbon.

These conditions are necessary for a large number of explosives and powders (Fig. 1). However, UFD yields significantly differ for various explosives. In general case diamonds yield is determined by the quantity of explosively obtaining condensed carbon (output CC - B_{cc}) and pressure value in chemical reaction zone (Pj).

The pressure role is clearly seen from the nature of the effect of charge density. Chapman-Jougue pressure is known to be proportional to cubic charge density. Thus, reduction in initial charge density causes a sudden drop of detonation pressure. Test results show that the reduction in charge density involves also a sudden drop of diamond yields (Table 2) (4).

The importance of the both mentioned factors (Bcc and Pj) is visually seen when analyzing UFD synthesis from mixture of TNT (trotyl) and RDX (hexogen) (6). Pure TNT gives the highest yield of CC; while adding RDX to it the yield of condensed carbon drops Dependence of CC yield on mixture composition has linear behaviour, showing that pure RDX is able to give about 2 % CC of ist own mass, and TNT-18,2 %. When RDX content is growing, the total power of detonation wave increases and pressure in it rises, favouring formation and growth of diamond form of carbon. Thus, in case of TNT-RDX system a competition of two processes take place. On the one hand, pressure growth jointly with increasing of RDX content facilitates synthesis of diamond particles, on the other hand the increasing RDX content reduces the amount of "building" material for their formation.

In the result, dependence of UFD yield (in % by mass of explosive - Bcc) upon TNT - RDX composition shows complex extreme behaviour (Fig. 2).

Maximum UFD yield is achieved with TNT content in the alloy is about 60 %. It should be noticed that similar relations for Bcc and B_{UFD} are also observed for TNT mixtures with other explosives: HMX, PETN etc. (5).

Fraction amount of diamond phase in CC presents agreat interest from both scientific and practical view of point. It is connected with the fact, that not the whole explosively obtained CC has diamond modification. Depending on charge composition and explosion conditions from 90 to 20 % of CC is isolated in non-diamand form (soot, graphite), which must be eliminated when obtaining pure ultrafine diamond.

Diamond phase dependence in CC on TNT - RDX composition (C_{UFD} in fig. 2) increases when RDX rises to 40 - 50 %. After that it becomes stable and remains constant when there is further increasing of RDX content. Such relation behaviour cannot be explained in terms of equilibrium representations about structure of chemical reaction zone of detonation wave. It needs more detailed investigation of UFD synthesis from composite heterogenous explosives using tracer atoms.

For this purpose it was used both TNT with C^{13}-tracer on methyl group (7) or C^{14} in benzene ring and wholly traced RDX (8).

Evaluted on the base of test results values of diamond amount, forming from RDX (D^{UFD}_{RDX} - the third column) and TNT (D^{UFD}_{TNT}) carbon are presented in Table 3 (8). It should be noticed that these quantities are determined on the base of independent experiments. In the fifth column the fraction which RDX carbon amounts for in a total carbon mass of composite explosive (m^{c}_{RDX}/m^{c}_{o}) is presented.

Received results have shown following:

1. Diamond phase when detonating TNT - RDX is formed mainly from the carbon being part of TNT, and UFD synthesis from RDX takes place to a lesser degree. Diamond fraction by weight formed from RDX is significantly less than contribution of this ingredient carbon in total carbon mass of composite explosives (see Table 3). So for TNT-RDX in the ratio of 80/20 (TNT - 80 %) only 2,6 % of UFD was formed from carbon atoms, entering into RDX in spite of that this carbon amounts for 9,9 % of its total quantity in charge. If during detonation fast products mixing occured in a chemical reaction zone and decay fragments of initial molecules statically occasionally reacted, as in the of ideal homogenous explosives, UFD fraction obtained from RDX carbon would be significantly larger. The result may be explained by unsufficient mixing of products of detonation of molecules invarious ingredients of composite explosives in a chemical reaction zone of detonation wave, where the formation of diamond particles takes place. This conclusion is confirmed by research of UFD synthesis from C^{13} TNT - RDX of various dispercity. Dispercity growth of this component reduces UFD fraction, formed from TNT, and increases UFD fraction from HMX, that is linked to increasing boundary surface area of ingredients and to the rise of mixing extent in detonation products of various ingredients of heterogenous mixture near the boundary.
2. All TNT atoms irrespective to its position in a molecule take part in UFD formation. It is confirmed by the fact, that the sum of diamond fractions, formed from TNT and RDX, is practically equal to 100 % (see Table 3) and also by close proximity of UFD fractions, formed from various carbon atoms of TNT molecule.
3. Results received by tracer method explain the observed relation behaviour of transition extent of CC to UFD. Transition extent relations of CC, separating from TNT (C^{TNT}_{UFD}) and RDX (C^{RDX}_{UFD}) into diamond phase, and total UFD fraction in CC are given in Fig.3. Relations are plotted assuming that RDX and TNT in mixture give such specific yield as when firing them in particular in pure form. This assumption is based on the conclusion that there is no appreciable mixing of detonation products of different ingredients of a composite heterogeous explosive in a chemical reaction zone. From Fig. 3 the transition extent of CC formed from TNT into diamond phase is seen to be

significantly composition of composite explosive it shows the following behaviour. At the beginning there observed a sudden growth which when TNT content in mixture is 60 % stops and the relation passes into plateau. Thus the obtained results show that RDX present in mixture with TNT increases the amounts of non-diamond carbon forms and total transition extent of CC into UFD lies in lower than the corresponding relation for TNT. Higher RDX content in the mixture the more CC "dilution" with non-diamond carbon. With low RDX content this dilution is balanced with sharp growth of transition extent of CC formed from TNT into diamond. When RDX extent is higher than 60 % the transition extent of CC from TNT into diamond terminates increasing the growth of non-diamond mass of carbon from RDX is not balanced with corresponding reduction for TNT. From this moment total dependence of transition extent terminates increasing and goes into plateau. Plateau existence on the relationship of carbon transition extent formed from TNT into diamond can be explained either by extent growth of TNT molecule destruction into fragments depending on power increase of detonation wave or by rise of formation velocity and diamond particles growth under these conditions. Under certain "threshold" wave power the effect of these becomes weak enough, the diamond formation velocity is maximum and further growth of wave power does not increase the velocity is maximum and further growth of wave power does not increase the velocity of UFD synthesis (6).

4. Significant difference in transition extents of CC, formed from TNT and RDX (see Fig. 3), into a diamond phase shows the importance of once more factor influencing on UFD synthesis of structural features of explosive molecule. This conclusion is confirmed by experimental results of various explosives when with proximate detonation parameters and CC yield the received diamond amounts are quite distinct.

Besides the above-mentioned factors, namely: CC yield, pressure in detonation wave, explosive structure, homogeneity of chemical reaction zone exerts significant effect on UFD yield from heterogenous mixtures increase of zone homogeneity causes to UFD yield growth. The relation of UFD yield is specific surface area of HMX (S_{HMX}) for TNT mixtures of this ingredient (see Fig. 4). HMX dispersity growth is seen to increase appreciably UFD yield. As average detonation parameters of the explosives don't depend on dispersity the availability of this influence confirms non-homogeneity of chemical reaction zone and existence in it microvolumes having different P,T - parameters and composition of DP. This influence is associated both with increase of participation extent of HMX carbon atoms in UFD formation with the growth of this ingredient dispersity and with rise of TNT fraction loaded with more powerful shock wave extending from HMX particles along surrounding TNT.

Considered factors concern the the zone of chemical reactions. Besides of them factors connected with charge detonation as a real physical body effect on UFD yield, namely : shape and geometry of charge, initiation power. The influence of given parameters is shown in Table 4, from which it is seen that the more powerful initiating shock wave the higher UFD yield. The same refers to the charge shape: elongated cylindric charge gives the more yield in comparison with plate charge. The influence of these parameters is connected with incomplete detonation of charge and UFD losses in subsurface fields and unsteady detonation zone near initiator. Besides the last investigations show that UFD yield must depend on the position of initiation point and the shape of initiation wave.

DIAMOND PRESERVATION IN THE PROCESS OF DETONATION PRODUCT (PD) SCATTERING IN EXPLOSION CHAMBER

One the most important problem is to preserve the diamond formed in chemical reaction zone of detonation wave while subsequent expanding PD, their mixing with surrounding medium and cooling. Depending on external conditions UFD yield from the same charge can achieve 8-9 % or fall off to zero. When charge fired in the air or in vaccuum free carbon practically disappears completely in detonation products (9). Influence of ambient conditions under which charge fire takes place on CC and UFD yields is connected with their effect on secondary reactions proceeding in PD while expanding and cooling.

The basic ambient factors effecting UFD yield while firing in chamber are the following: pressure and gas nature, filling the chamber before explosion; charge weight; volume of explosion chamber. The effects on UFD yield. With other conditions being equal UFD output increases:

a) with pressure rise of the gas filling the chamber;
b) when using gas with more thermal capacity and molecular weight;
c) with reduction in charge weight;
d) with increase of explosion chamber volume.

The obtained relationships can be explained if take into account that external conditions practically have no effect on character of detonation wave spreading. But they exert an essential influence on the flow behaviour of PD in chamber after explosion has been carried out in one dimensional approximation.

The result have shown the following picture of PD movement. After charge explosion detonation products begin to scatter from the surface into the surrounding medium. In surrounding gas there is a shock wave moving towards walls of chamber. Their direct shock wave reaches the chamber walls, reflects from them and starts moving to meet expending PD. At some moment they encounter after which the explosion products begin to contract themselves and the reflected shock wave "collapses" in the centre of chamber and reflects from it. Further, the described processes and repeated with low intensity.

Modeling of PD movement shows that there are several stages during their movement, which are characterized by the essential temperature increase, acceleration of secondary reactions, proceeding in products after explosion and that makes it possible some losses of the formed UFD.

They are initial stage of PD expansion, stage of internal shock wave and reflected shock waves. Comparison of modeling results with experimental data permits to conclude that the largest UFD losses occur in the first reflected shock wave losses at other stages are much lower. There has been observed one-valued correlation between the temperature of detonation products in the reflected shock wave and the UFD output.

Nitrogen pressure influence upon the detonation products temperature in the reflected shock wave as shown in Fig. 5. Nitrogen pressure rise is seen at the range from 1 to 8 atm and temperature of PD primary mass reduces of more than 400 degrees. Experimental data shows that it leads to output increasing of the UFD from 0.9 up to 6.8 % (see table 5).

Nitrogen substitution for CO_2 (at 1 atm) causes temperature reduction of PD up to 250 degrees (Fig. 6) and increases output up to 3.5 % (Fig. 5). Charge weight and volume of explosion chamber exert similar effects.

Assuming that the main loss of UFD in reflected waves are caused by graphitization, having zero order concentration, and supposing that at the first approximation the reaction runs within the narrow temperature range proportional to internal energy of PD it is shown that high precision output of UFD from various explosives can be described by means of common expression:

$$B_{UFD} = B_{UFD}^{MAX}(1 - 11{,}473\,EXP(9029{,}8 / E_{PD}^{MAX}), \qquad (1)$$

where B_{UFD}^{MAX} - maximum UFD output, that is diamond amount which detonation products contain after initial state of their scattering, and E_{PD}^{MAX} - internal energy of PD when reflected shock wave is maximum.

Plot of the received relation and relation and experimental points for various explosives are given in Fig. 7.

Good agreement of the resulting relation with experimental points for different explosives shows that diamond particles formation and their loss during the charge explosion in the chamber take place at different independent from each other stages:

- UFD synthesis in a zone of chemical reaction of detonation wave;
- UFD loss in reflected shock wave.

Besides, B_{UFD} term depending on (1) gives UFD output at synthesis, expression in brackets describes UFD loss. The factors such as composition, and structure of explosives, geometry, shape and homogeneity of charge, power of initiating shock wave exert primary influence on B_{UFD}^{MAX} charge weight, nature and pressure of chamber medium before explosion, volume of explosion chamber exerts influence on the diamond loss level in secondary reactions.

Investigation of physical chemical relationships of synthesis and UFD preservation, study of their properties allowed to realize manufacture of UFD in production scale.
Block diagram is presented in Fig. 8.

Process of production includes the following operations:

- development of chamber medium necessary for synthesis;
- explosion of the charge;
- trapping of the condensed products of explosion;
- oxidation of non-diamond carbon forms;
- UFD purification from metal impurities;
- washing and UFD extraction from suspension;
- packing and production.

PHYSICAL CHEMICAL PROPERTIES OF UFD

The formed UFD according to the conditions of synthesis has on its surface non-diamond forms of carbon, sorbated and chemosorbated groups of the carboxile, hydroxile and methyl constractions.

The chemical purification of non-diamond forms of carbon keeps 4-5 % of chemosorbated groups on the UFD surface. While heating, their concentration decreases and at the temperature of 973 K, they cannot be found by the method of polarography (10).

The purified UFD is the ultra-dispersed powder with the specific area of surface 280...350 m^2/g, the volume of pores 0,3-1,0 cm^3/g and the density 3,1-3,2 g/cm^3. The size of micro-chrystallites, according to the data, obtained by the method of small-angle dispersion of X-ray, is 3-6 nm, the form of particles being close to spherical. The morphology and the size of UFD particles let us call it the cluster UFD.

In accordance with the elements analysis data, the cluster UFD contains carbon (up to 89 %), hydrogen (up to 1.5 %), nitrogen (up to 4.5 %) and oxygen (up to 10 %). At thermotreatment in reduction and neutral medium the content of carbon in samples is within the maximum range of 973-1073 K, while that of hydrogen and nitrogen is actually constant.

The cluster UFD has the cubic crystal lattice with the parameter of 0.3562 ± 0.003 nm, the size of regions of coherent dispersion of 4-6 nm, 1 % of the microstress of II type (10 GPa). This witnesses the highly defective and highly deformed crystal lattice of the cluster UFD, it is thermally stable as the value of microstresses do not vary at heating actually up to the temperature of the start of graphitization (1473 K). The high stability of UFD to baking not with standing, its reactionability is really high.

Oxidation in the air atmosphere at 700 K, while artificial diamond of static synthesis is oxidized at 860-920 K.

At room temperature the UFD resolves water solutions K_2 Pd $C1_4$, Fe $(C_2H_3O_2)_2$, with extraction of pure metal, which coordinates with changes in the chemical potential of system.

The cluster UFD possesses a high absorption capacity , which should be reflected upon the characteristics of the systems, containing it. Thus, by the example of water solution of KC1 it is obvious that one volume of UFD may lattice - order up to 300 volumes of liquid. As a result, it is impossible to introduce more than 3-6 % volumes of these powders to the liquid.

These effects are studied in more details in metals. Electrochemical and chemical precipitation of metal on the surface of the metal matrix of electrolytes in presence of UFD leads to the changes in size of the crystals of precipitated metal, and those of its surface. Considerable increase of the microhardness and wear stability is observed simultaneously. According to the data of investigations, the granule of deposit on the basis of chrome decreases 1.4 times, the size with microhardness being increased 1.25 times as much; and the size of crystallites deposits on the basis of nickel and copper becomes 10 times less while microhardness increases 1.5-3.5 times as much.

Measuring of the relative intensity of reflection from various crystallographic surfaces of the film of precipitated electrochemical chrome has showed that presence of UFD in the electrolyte leads to disorientation of crystallites in the surface of precipitation, and to distortion of the chrome lattice parameter.

This means that in the process of forming of composite coverings on the basis of various mettal matrices, considerable change of the metal phase structure occurs. For this resason, and because of the presence of UFD in the structure of coverings, they have much more high wear stability as compared with regular coverings (table 6).

At the present moment, the technology of making chrome-diamond coatings is put into practice in more than 100 large enterprises. The obtained results have showed considerable increase of stability of the metal-cutting instruments (table 7).

The UFD or its mixture with non-diamond forms of carbon UFGD (UFGD-ultrafine graphite-containing diamond) may also be introduced to melting metal, and the same effect of increasing of microhardness is evident too (table 8). Characteristics of rubbers (table 9) and polymers change (table 10).

All the witnesses the unique nature of influence of the cluster UFD on materials, and is connected with its instrumental properties. It is also interesting to study another region of its application, which is connected with the small size, high absorption capacity and roud form of particles. Introduction of such diamond to oil leads to temperature decrease in friction area, decrease of friction coefficient in roller bearings 1.5-1.8 times as much and that of wearing time of the rubbing pair 2-10 times as much; this allows to increase limit loading on the friction assembly 1.5-7 times as much. Because of the selective adsorption oil viscosity decreases, while breaking and destruction of the oil film ("dry friction") occur at much higher ranges of relative gliding. This guarantees the best properties of oils (11).

Studies of the Tokyo University Yamanassi (Japan) have shown that introduction of UFGD to the oil 220 by ICO leads to decreasing of the friction coefficient for the contact pair of gliding 10 times as much. The Research Institute of Oil Industry (Hungary) records the fact that introduction of UFGD to oils manufactured in Hungary DS-5 (motor) and NUKOMOL K 85 W-85 W-90 (transmission) improves their antifriction and operating properties.

In practice oil additives on the basis of UFD have been tested in industrial companies AVTOVAZ, KAMAZ and are successfully applied to the internal combustion engines (both to carburator and diesel ones). In the KAMAZ engines the economy of lubricant reaches 25 %, fuel - up to 8 % with simultaneous decreasing of run before reaching the maximum characteristics in the period of exploitation car driving 2-3 times as much. In the AVTOVAZ engines introduction of the additive to motor oils enables to increase the process of grinding of new engines and those after thorough repair 2-3 times as fast, and to decrease the average expenditure of fuel per hour 2-3 % as much.

The additives have been successfuly applied to the precision metal processing tools in various countries.

Increase of smoothness of run, keeping of high accuracy of treatment when regular unalloyed industrial oils are used in spite of special ones, decrease of expenditure of oils 2-3 times as much are evident.

The same picture is also obvious for the plastic and solid lubricants based on the UFD.

The cluster diamond is a new product and some of its useful properties are not investigated sufficiently enough to determine new regions of its application. However, the data mentioned above let us conclude that application of the cluster diamond is really perspective and researches in this direction are constantly widening at present.

REFERENCES

1. A. M. Staver, N.V. Gubareva, A. I. Lyamkin, E. A. Petrov. Physics of burning and explosion, 1984, v. 20, No. 5, p. 100.
2. A. I. Lyamkin, E. A. Petrov, A. L. Yershov et al. Reports of the USSR Academy of sciences, 1988, v. 302, No. 3, p. 611.
3. N. R. Greiner, D. S. Phillips, J. O. Johnson et al. Nature, 1988, 333, No. 6172, p. 440.
4. V. Sakovich, V. M. Titov, P. M. Bryliakov et al. X Intern. Conf. HERF.-1989, Ljubliana, Yugoslavia. Proceedings, p. 179.
5. V. F. Anisichkin, I. Yu. Malkov, F. A. Sagdeyev. V All-Union Conference on detination. Reports.-1991, Krasnoyarsk, p. 27.
6. N. V. Kozyrev, E. S. Golubeva, Physics of burning and explosion, 1992, v. 28No. 5.
7. V. F. Anisichkin, B. G. Derendiayev et al. Reports of the USSR Academy of sciences, 1990. V. 314, No. 4, p. 879.
8. N. V. Kozyrev, P. M. Bryliakov et al. Reports of the USSR Academy of scienes. 1990. V. 314, No. 4, p. 889.
9. E. A. Petrov, G. V. Sakovich, P. M. Bryliakov. Reports of the USSR Academy of sciences, 1990, v. 313, No. 4, P. 862.
10. L. A. Petrova, A. L. Vereschagin ei al. Superhard materials, 1989, No. 4, p. 3.
11. G. V. Sakovich, P. M. Bryliakov et al. JVHO named after D. I. Mendeleyev 1990, v. 35, No. 5, p. 600.

Table No. 1
UFD synthesis from several explosives

E	ρ_o, g/cm^3	Pj K	B_{cc} %	B_{UFD} %
TNT	1,59	17,6	18,7	2,4
TNT/RDX-50/50	1,68	24,2	10,2	6,8
TNT/HMX=50/50	1,72	30,0	9,8	6,6
TNT/PETN=50/50	1,66	26,6	8,2	4,8
TNT/BTF=50/50	1,68	26,2	11,3	7,1
Tetryl	1,61	22,5	9,1	4,1
DINA	1,62	28,0	3,0	0,8
TNT/DINA-50/50	1,58	23,9	9,0	3,1
RDX/paraffin 78/22	1,45	19,2	18,4	2,3
RSI-60 (ballistic)	1,59	19,6	9,5	2,9
RNDSI-5K (ballistic)	1,61	18,6	8,5	2,0

Table No. 2
Influence of charge density upon the output of CC and UFD from the composition of TNT/RDX-50/50.

ρ_o, g/cm^3	B_{cc}, %	B_{UFD}, %	C_{UFD}, %
0,90	4,3	0,96	22,3
1,60	10,3	3,99	38,7
1,68	10,2	6,82	67,18

Table No. 3
Results of experiments with labelled TNT and RDX

HE	ρ_o g / cm^3	D^{UFD}_{RDX} %	D^{UFD}_{TNT} %	m^c_{RDX} / m^c_o %
TH - 80	1,64	2,6	97,3	9,9
TH - 70	1,65	4,4	—	15,8
TH - 60	1,67	4,9	94,7	22,6
TH - 40	1,70	14,0	84,0	39,7

Table No. 4
Influence of the charge geometry and the initiation power upon the UFD output from the mixture TNT/RDX - 60/40

Diameter of charge mm	Initiator	B_{cc} %	B_{UFD} %
50	8g RDX + ED*- 8	11,8	6,8
38	8g RDX + ED-8	11,7	7,8
38	ED-8	11,3	6,0
25	8g RDX + ED-8	11,4	8,1
25	ED-8	11,6	7,2

* ED - electric detonator

Table No. 5
Influence of external conditions upon the output of CC, UFD from the mixture TNT/RDX - 50/50

Volume chamber, m^3	Medium		Charge		B_{cc} %	B_{cc} %	Notes
	Caz	ρ_o,MPa	m_{HE}, kg	ρ_o, g/cm^3			
0,175	N_2	0,1	0,1	1,68	5,4	5,4	Influence of ρ_o
the same	N_2	0,3	0,1	1,68	10,1	5,2	
the same	N_2	0,8	0,1	1,68	10,2	6,8	
the same	CO_2	0,1	0,1	1,66	10,9	3,6	Influence of gaz
the same	He	0,1	0,1	1,68	3,7	0,08	
3,0	PD	0,1	0,65	1,64	10,4	5,1	Influence of charge mass
the same	PD	0,1	0,83	1,64	7,8	3,1	
the same	PD	0,1	1,5	1,64	5,0	1,9	

Table No. 6
Influence of UFD on microhardness and weariness of metal

	Microhardness HV, kg/mm^2		Decrease of weariness
	without UFD	with UFD	
Nickel electro-chemical	350-450	625-725	3-4
Nickel-cobalt electrochemical	550	885	1.6-2
Nickel chemical	525	725	1.4
Copper electro-chemical	170	550-600	3.5
Chrome electro-chemical	900-1100	1100-1400	1.5-2

Table No. 7

Prolongation of life of instruments with chrome-diamond coatings

Instruments	Prolongation, times
Matrices for cool stretching of metals	2...5
Press-instrument for powder metallurgy	over 10
Stamping instruments	1.4...4
Steel drills	8...12
Cutting instruments: for metal	1.5...4
for glasplastics	3...10
Shafts, gears of tools and mechanisms	2...3

Table No. 8

Characteristics of the cakes of UFD and aluminium

Composition	Hardness by Brin. N/mm^2	Microhardness kg/mm^2	Density g/cm^3	Heat conductivity V/m, K	Porosity %
0 % UFD	30	78	2.73	162	0
2 % UFD	82	181	2.71	141	0.5
10 % UFD	95	197	2.69	101	1.05
20 % UFD	117	268	2.71	64	1.9
50 % UFD	-	522	2,6	9,2	1.5

Table No. 9

Relative growth of indices of rubber with UFD

Properties of rubbers	Rubbers on the basis of polar caoutchoucs	Rubbers on the basis of non-polar caoutchoucs
Wear resistance	1.1...3.2	1.1...5.3
Stability to repeated stretching deformations	1.1...1.2	1.1...1.6
Tear resistance	1.05...1.4	1.0...1.3
Resistance to increasing of chaps		1.1...1.3
Thermostability in stability	1.0...1.1	1.0...1.15
Resistance to thermal aging:		
in stability	1.05...1.1	1.0...1.15
in elongation	1.03...1.1	1.05...1.15

Note: Other properties of rubbers are the same

Table No. 10
Results of comparative experiments of polytetrafluorine-ethylene (PTFE) with UFD additives

Material	Module of elasticity at pressing, MPa	Friction coefficient	Relative wear-stability
PTFE	4.2 ± 0.1	0.21	1
PTFE + 20 % coke (F_4K_{20})	12.2 ± 0.5	0.32	25
PTFE + 5 % UFGD	12.2 ± 0.5	0.21	68

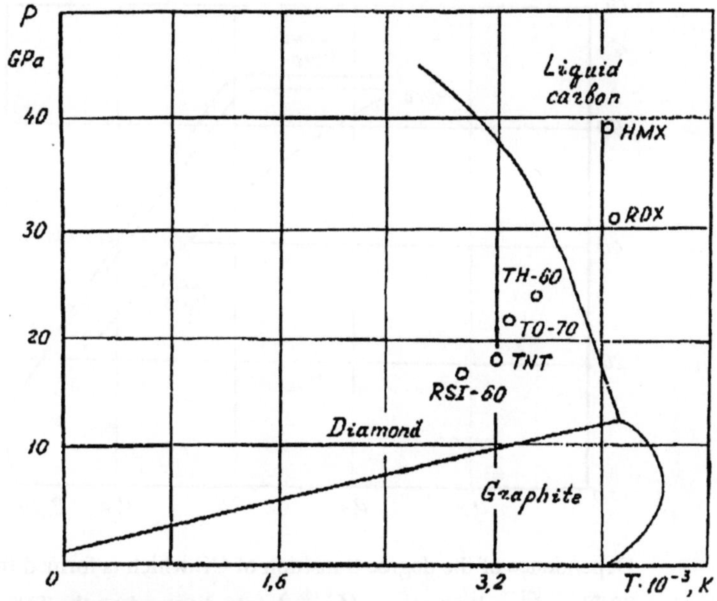

Fig. 1 Phase diagram of carbon

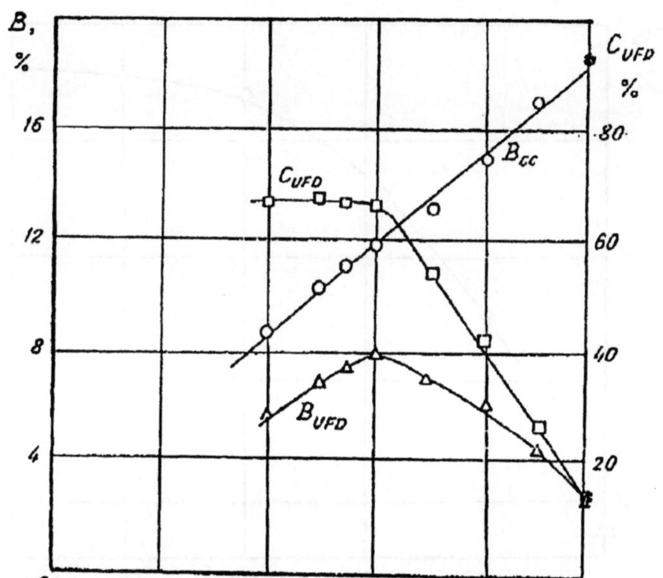

Fig. 2 Output dependence of CC, UFD and diamonds in the condensed carbon on the TNT content in the TH mixture

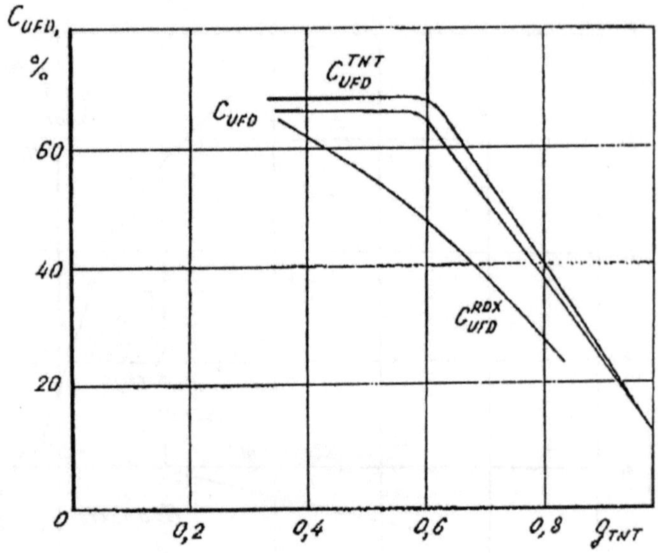

Fig. 3 Dependence of the degree transition of CC which is formed from TNT (C^{tnt}_{ufd}) hexogen, (C^{rdx}_{ufd}) into diamand on the TNT content in the TH mixture

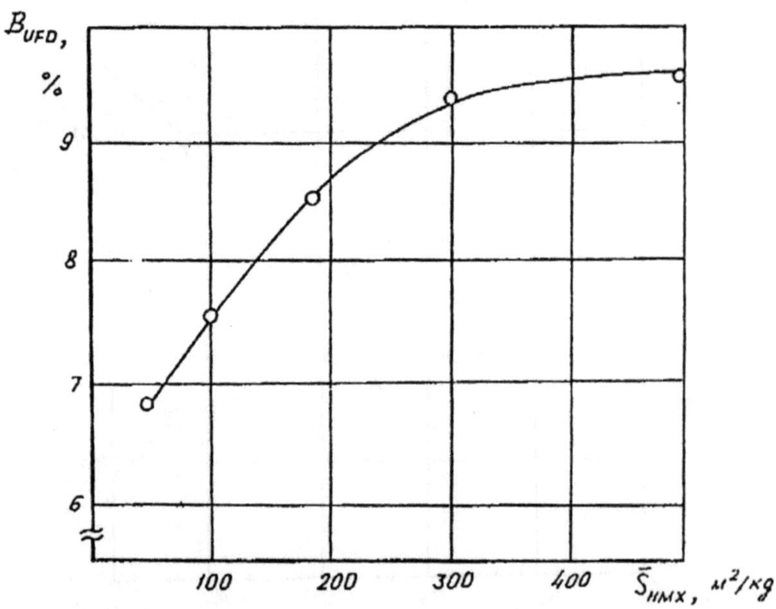

Fig. 4 Influence of octagen dispersity in the mixture TNT/HMX - 70/30 on the UFD output

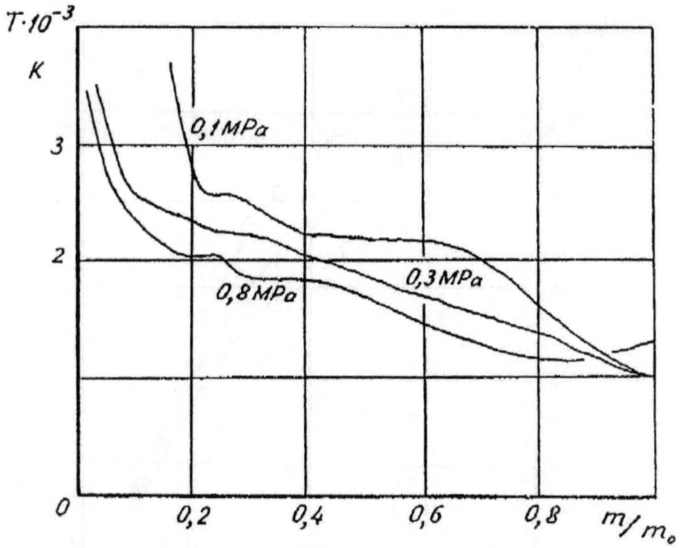

Fig. 5 **Influence of the nitrogen pressure upon the temperature of PD in the reflected shock wave.**
(Vc = 0,175, m_{he} = 0,1 kg)

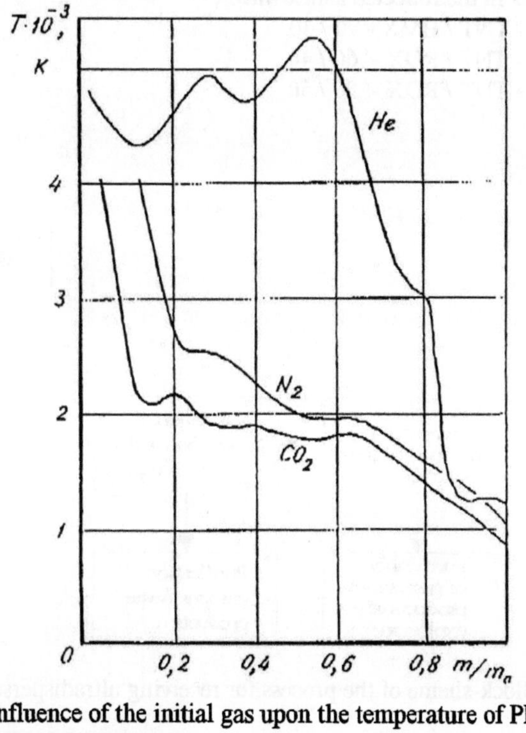

Fig. 6 **Influence of the initial gas upon the temperature of PD in the reflected shock wave.**
(Vc = 0,175 m3, m_{hee} = 0,1 kg, Po = 0,1 MPa)

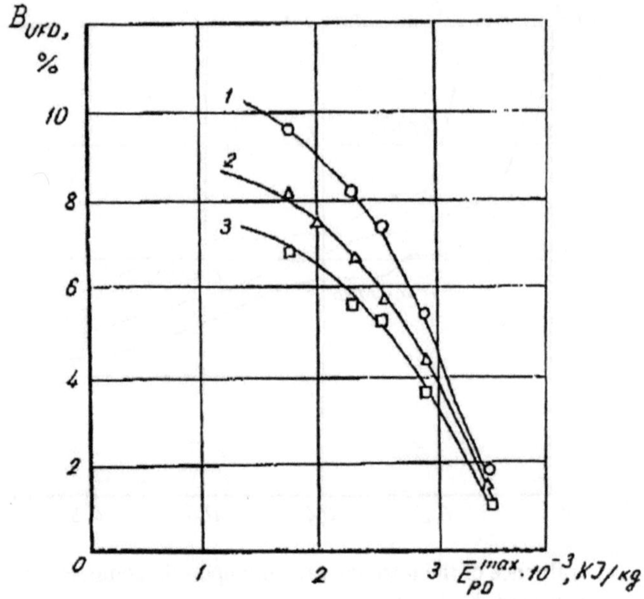

Fig. 7 Dependence of the UFD output on the maximum interior energy of PD in the reflected shock wave.
1 - TNT / HMX - 70 / 30
2 - TNT / RDX - 60 / 40
3 - TNT / RDX - 50 / 50

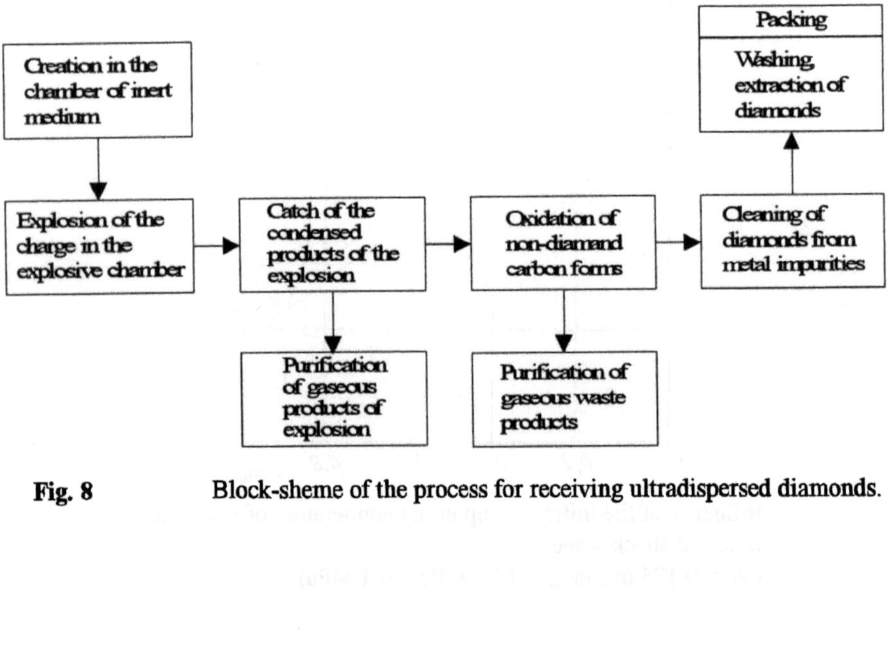

Fig. 8 Block-sheme of the process for receiving ultradispersed diamonds.

COMPREHENSIVE UTILIZATION OF CONVENTIONAL AMMUNITION AND RELEASED EXPLOSIVES

B. V. Matseyevich and Yu. M. Leshchinsky

Recent general changes of the international situation and conclusion of a number of international treaties on weapon and war materiel reduction have made topical the problems connected with utilization not only of nuclear and chemical ammunition (weapon of mass destruction) but also of conventional ammunition: artillery, aviation, naval, etc.

In the years of the "cold war" many countries including the Soviet Union have accumulated substantial stores of conventional ammunition storing of which is not only very costly but also presents potential and evergrowing risk of mass explosions, fires, and other ecological catastrophes.

At present, a large quantity of ammunition to be utilized is accumulated on the Russia's territory.

Such an ammunition, as a rule, was exploded or burned up on the open sites or sunk in the depth of the world ocean.

There is no need to comment the ecological damage from such methods of demilitarization and, beside that, a substantial quantity of material resources is irretrievably lost.

Introduction of special legal regulations on enviromental protection and Greenpeace movement in many countries of the world have blocked the use of polluting methods of demilitarization and stimulated search for unpolluting and cost effective methods of utilization.

Taking into account the large volumes of ammunition to be utilized on the territory of Russia, the possibility to use detonation, burning up or sinking in the depth of the world ocean is completely excluded. That is why, the utilization methods are chosen as a general approach to this problem in this country.

The approach to the problems of utilization of extra stocks of ammunition in the Russian

H. Krause (ed.),
Conversion Concepts for Commercial Applications and Disposal Technologies of Energetic Systems, 73–78.

Federation is based on the following main principles:

- comprehensive utilization of all ammunition elements;
- safety at all stages of the work;
- entirely unpolluting technological process;
- registration of ammunition, its elements and explosives which are the products of utilization in view of their extreme social danger and taking measures against their unauthorized spreading;
- creation of new jobs;
- economical expediency when chosing this or that method of utilization.

Scientific and technical problems in constructing of comprehensive facilities for ammunition utilization are connected with:

- explosives extracting from the shell bodies;
- reprocessing of the extracted explosives into industrial explosives;
- reprocessing of powders both into industrial explosives and into other products and articles;
- constructing of special furnaces for unpolluting incineration of detonating fuses, primer sleeves, tracers, etc.;
- reprocessing of packing and other auxiliary materials into commercial products;
- observance of environmental protection requirements under utilization.

The comprehensive utilization of storing on the bases and arsenals of the Ministry of Defence of Russia stocks of outdated and unserviceable ammunition will yield the following rensable materials:

- ferrous metals
- non-ferrous metals
- explosives
- powders and solid propellants
- wood.

It is absolutely obvious that solution of the entire complex of problems connected with the ammunition utilization will require not only substantial financial expenses but a lot of time.

Meanwhile, the practical utilization cannot wait. That was why already in 1992 some plants of Russia began the utilization.

To eliminate or reduce the ecological damage before some scientific, technical, financial and organizing problems are solved, at present, only the most simple in design ammunition filled predominantly with trotyl (artillery common shells, antitank mines and some others) are utilized.

Such elements as detonation fuses, primer sleeves, for which available methods of utilization are connected with large ecological damage, are stored in awaiting of dependable equipment for their reprocessing.

One of the main problems of utilization is extraction of explosives from the ammunition. At present, there is no universal method of ammunition disposal. It is connected with variety of both ammunition constructions and explosive compositions used for filling, having wide range of physical and mechanical or physical and chemical properties.

For every specific type of ammunition specific technology of explosives extracting and reprocessing should be developed taking into account economical expediency and desired level of environmental protection.

These technologies may be conventionally united into three groups:

1. Different methods of noncontact and contact melting out by means of steam or hot water, use of induction heating of the shell body, washing out by melted trotyl for ammunition filled with trotyl or other fusible explosives on the base of trotyl.
2. Different methods of washing out by high-pressure water jets are advisable for large ammunition filled with fusible composite propellants.
3. Different methods of mechanical destruction of explosive charges including destruction by water jet for ammunition filled, mainly, with non-fusible explosives of A-IX-1 and A-IX-2 types pressed into the shell bodies.

The extraction of explosives (explosive charges) from ammunition loaded with separate inserted explosive blocks with fusible fixative produces no difficulties. On healting such shell bodies the fixative melts and the explosive charge is readily extracted.

Especially important for the entire problem of utilization is to find the fields of application for explosives and powders extracted from the ammunition as its destruction by exploding or burning brings the most damage for the nature.

Analysis of ammunition to be utilized shows that it contains of trotyl compositions: TA (trotyl/aluminium) and TD (trotyl/dinitronaphtalene), of fusible mixtures: (trotyl/hexogen) or (trotyl/hexogen/aluminium), of non-fusible explosives: A-IX-1 (phlegmatized hexogen) and A-IX-2 (mixture of phlegmatized hexogen and aluminium), of other explosives including liquid and plastic explosives, octogen - containing and some other explosives.

If compare above quantities with demands of the mining industry (Russia produced about 1.3 million tons of industrial explosives in 1992), the use of explosives extracted from ammunition will make no substantial influence on the total output because these explosives cover only 2-3 % of the yearly demand, but their absolute quantity has large material value and also must not be destructed by ecological considerations.

At present, the extractive branches of industry use mainly explosive compositions on the base of trotyl and ammonium nitrate. As a rule, they are characterized by low rate of detonation and low sensitivity to shock and friction. The rock demolistion by industrial explosives containing some quantity of hexogen has practically stopped by now because of high danger of their production and application. Development prospects of industrial explosives are oriented on wide use of a new class emulsion explosives.

The mining industry has practically no need in hexogen - containing explosives because of their high sensitivity to shock and friction. The entire technology of blasting operatings, mainly, is not adapted for use of such explosives.

Of all explosives containing in the ammunition only trotyl and composition "trotyl/aluminium" are dual-application substances used both for military and civic purposes, but even these explosives after melting out or other extracting from ammunition must be reprocessed to use in the mining industry.

The main method of reprocessing is granulation. For this purpose belt granulator with special disperser forming melted drops of explosive is developed. Directly after granulation, explosives may be delivered to consumers.

Mixtures "trotyl/hexogen" and "trotyl/hexogen/aluminium" extracted from the ammunition must be adjusted to reduce their sensitivity to mechanical disturbances and for universal application in the mining industry. That is achieved both by increasing the content of trotyl in the compositions and introducing some inert components.

The great efforts are in view not only in developing of new unified compositions of industrial explosives but also articles from those explosives with sufficient volume of the governmental tests. To this purpose, technology for granulation of such compositions as trotyl/hexogen, trotyl/aluminium, trotyl/hexogen/aluminium including ones with ammonium nitrate additive is successfully developed by now.

It is necessary to define and legalize operating conditions for these compositions.

The next task is the use of extracted from ammunition high-strength explosives consisting of phlegmatized hexogen or mixture of phlegmatized hexogen and aluminium and their reprocessing into industrial explosives. Since they do not melt, the most perspective way is manufacturing on their base of slurry-type charges.

Such charges are perspective especially for users with small and medium explosives consumption, and growth of the volume of their production may be expected in the nearest future. It is no doubt that they will find wide application in the mining industry of Russia.

It is absolutely obvious that reprocessing of explosives, especially hexogen-containing ones, for further use in the extractive branches of industry is not cost effective. At the same time, total ecological damage from burning-up of extracted explosives without their reprocessing into industrial explosives would make 60 billion rubles loss in prices on 1-st January 1993. All expenses for their utilization will not exceed 30 billion rubles. Even these figures say that we are doomed to look for ways of explosives reprocessing.

It is obvious, that the main direction of application of extracted explosives as industrial explosives and different articles (relay detonators, extended jet steel cutting charges, cable charges for forest fire extinguishing by forming of mineralized strips, special charges for demolishing of oversized stones, for hardening and welding of metal structures) is not the only possibility.

Possible perspective ways of application of extracted explosives are explosive production of technical diamond, pigments production, production of triaminotoluene from trotyl as a feedstock for chemical industry, etc.

These ways are described in special reports of different enterprises which work in accordance with agreed technical asignments

Institutes of the Russian Academy of Sciences, spezialized scientific institutions and sub-faculties of institutes of Higher Education are involved in this work.

The Russian Federation is forming a comprehensive approach to the problem of utilization of conventional ammunition and explosives which gives confidence that this difficult from scientific, engineering and production point of view problem will be solved.

SELF-PROPAGATING HIGH-TEMPERATURE SYNTHESIS: HISTORY OF ONE INITIATED CONVERSION

A. G. Merzhanov

Abstract

In the beginning of the 60s, the author's laboratory (at the Branch of the Institute of Chemical Physics, Chernogolovka) was engaged in the studies on the mechanism of combustion in gasifying condensed systems such as explosives, gun powders, and solid propellants. These studies were based on the concepts formulated by A.F. Belyaev, Ya.B. Zel'dovich, B.P. Zhukov, P.F. Pokhil, and other well-known researchers in the field of combustion. These studies aimed at elucidating the combustion patterns in modern installation (combustion chambers) and were oriented on the needs of arms industry.

In these studies, we tried to demonstrate that the combustion patterns in these systems were closely related to the gasification processes (gas evolution in reactions taking place in condensed phase and disintegration of starting reactants giving gas suspension). It was decided to search for a model system, in which gasification would be absent and all of the compounds involved would be present in their solid state. Such a system was constructed on the the basis of the iron-aluminium thermite diluted with alumina. The conclusions did not make them wait -- the data obtained were consistent with simple theoretical predictions and confirmed the expectation that the specific features of these systems are closely related to the transition of starting condensed reactants to the gaseous combustion products.

The simplicity of the model system fascinated the researchers. A desire appeared to seek for other systems of this kind. A comprehensive search has lead to a new class of combustibles: it were the powdered mixtures of metals (titanium, zirconium, hafnium, niobium, etc.) with nonmetals (boron, carbon, silicon, etc.). Due to high caloricity of the systems, the travelling front of oxygen-free combustion was formed, upon local initiation

H. Krause (ed.),
Conversion Concepts for Commercial Applications and Disposal Technologies of Energetic Systems, 79–84.

and a starting mixture was converted to an incandescent but condensed combustion product.

In the early experiments, our attention was drawn to the fact that the combustion products comprised the valuable refractory compounds (borides, carbides, silicides, etc.), the synthesis of which was attended by serious difficulties both in laboratory and industry. As a result, a new efficient method for manufacturing refractory materials -- Self-propagating High-temperature Synthesis (SHS) -- has been developed. The development of SHS can be regarded as a conversion project, because here the use was made of the experience and knowledge acquired in the course of defence-oriented studies to formulate a new trend in the chemistry and technology of inorganic materials.

Much attention was paid to the R&D that ensured the development from the formulation of fundamentals to setting-up industrial production, including chemical synthesis and the development of new technologies, materials and instrumentation.

Fundamentals of SHS

SHS investigations aim at elucidating the wave propagation patterns, the zone structure of a combustion wave, the chemical, physicochemical, and fluid dynamics phenomena in the wave, the heat balance in the system, the phase and structure transformations in combustion products. These studies are carried out using the modern methods of experimental diagnostics, combustion theory, kinetics, chemical thermodynamics, and high-temperature physical chemistry. The data of comprehensive studies allowed formulating the control means for burning velocity, temperature and degree of conversion; whereas the accumulated experience ensured reliable controlling of the composition, morphology, and structure of end products.

The SHS studies gave rise to a new branch of fundamental research -- structural macrokinetics. This discipline treats the joint occurrence of structural and phase transitions, chemical reactions and heat/mass exchange with special attention to positive and negative feedback between them.

The SHS processes are most comprehensively studied for the powdered mixtures of metals (Ti, Zr, Hf, V, Nb, Ta, Cr, Mo, W, etc.) with nonmetals (B, C, Si, S); for the intermetallic mixtures of Ni, Co, Ti, Al, etc.; for the hybrid systems of the metal-gas type with H_2, N_2, O_2, as reactants, and for the metallothermic mixtures *metal oxide--nonmetal oxide (or nonmetal) -- reducing metal (Al, Mg).*

To date, the attention of researchers is focused on the multicomponent systems that offer a variety of transformation pathways in a combustion wave.

Chemical Synthesis

Chemical synthesis is a main designation of SHS. Combustion is carried out in a way that ensures obtaining an end product of a desired chemical and phase composition. In synthesizing chemical compounds, it is desirable to obtain a single-phase product of minimum contamination. This is achieved by a prober selection of a green mixture, by ensuring complete combustion and sufficient exposure at high temperatures.

By the SHS method, the following systems are being produced: binary compounds and solid solutions of chemical elements, solid solutions of binary compounds, multicomponent solid solutions and compounds, nonstoichiometric phases, high-temperature and high-pressure phases, etc.

Chemical synthesis is a fine art that requires not only a deep insight into the mechanism of SHS but also a good deal of intuition.

To date, about 500 various compounds have been synthesized by SHS: refractory compounds (borides, carbides, nitrides, silicides), oxides (tanthalates, niobates, ferrites, cuprates, etc.), intermetallics (aluminides, germanides, nickelides), chalcogenides (sulfides, selenides, tellurides), phosphides, hydrides, and some others. Their number has no principal limitation. A general trend in the field can be characterized by a gradual refusal from the search for new synthetic capabilities of SHS in favor of specific purpose-oriented synthetic problems.

SHS Materials

Chemical synthesis is closely related to the problem of developing new advanced materials. In this respect, SHS also demonstrates its capabilities, because the products obtained can acquire not only a desired composition but also structure, thus providing the control means for their physical and service parameters.

To date, about 100 projects have been developed for manufacturing the materials of varied designation. Among them are hard alloys, cutting materials, nitrided ferroalloys and alloying agents, advanced ceramics, refractories, heat resitant and temperature-strong intermetallics, light alloys, superconducting ceramics, solid lubricants, and many others.

On the basis of SHS, new advanced materials are being developed: composite powders, foam ceramics, ceramics containing no activators of sintering, anisotropic ceramics, and oxygen-free single crystals. Special attention is paid to the development of macroinhomogeneous structures, i.e., the materials with an intentionally organized inhomogeneous structure: coated materials, multilayered and functionally-gradient materials, composites, welded workpieces, etc.

SHS Technology

The special capabilities of SHS best manifest themselves in the applications of SHS. Such outstanding features of SHS as the internal heat release and high burning velocity allow highly-productive and resource-saving manufacturing new high-quality materials to be suggested.

To date, about 30 SHS production methods have been suggested. They can be classified in six technological types. They differ in the composition and structure of a green mixture, in the burning conditions, in the composition and pressure of gaseous environment, in the type and intensity of external action, and in their morphology and destination of end products.

Two approaches are being used in SHS technology. The first of them is the production of not end but intermediate products, which then are used as raw materials in further processing. In this case, synthesis, structuring, and shaping are carried out in a one-stage process.

The facilities of SHS technology are more simple than that used, e.g., in powder metallurgy (because heating elements are absent). Among SHS facilities, the universal reactors with 20-30 l in volume became most popular (SHS-20 and SHS-30). These reactors are used to manufacture the elemental, magnesiothermic, and aluminothermic powders, to perform SHS surfacing and SHS sintering. The list of SHS facilities available includes SHS cameras with thermovaccum treatment of samples, SHS gasostats of high pressure, special presses equipped with press dies of varied size and shape, nondestructive explosion vessels, SHS extruders and roll mills, equipment for SHS forging, centrifugal SHS machines, apparatures for SHS welding, etc.

In Spite of considerable progress in SHS technologies, their technical level should be regarded as the pilot-scale one. The productivity of SHS installation is sufficiently high. However, a new level of productivity can be attained upon further improvement in the instrumentation and setting-up the automated in-line production. Among the works in this direction, the Spanish-American-Russian „Prometheus" Project should be mentioned: this project aims at setting-up a modern SHS production line for powders on an industrial scale (F. Llorente, M. Werle, I. Borovinskaya).

Some SHS technologies have found their industrial applications. In Russia, the USA, and Japan, the SHS powders of inorganic compounds, nitrided ferroalloys, ceramic insulators, high-temperature heaters, shape memory alloy wire, cutting inserts, ferrite articles, etc. are produced on an industrial scale.

Technical and Economic Efficiency of SHS

In view of actual penetration of SHS into industry and engineering, much attention is being paid to the studies on the efficiency of SHS. The following physical factors were found to define the efficiency of SHS: high temperature and high burning velocity (high degree of conversion and self-purification ensure high quality of products; high productivity), internal heat release (energy saving, the feasibility of dealing with large amounts of green mixture and producing large-scale articles, simplicity of installation), controlled cooling rate (autoannealing and autoquenching, controlling the material strength).

The technical efficiency of SHS is normally determined by comparing the service parameters of SHS products with that of conventionally obtained products taken as a prototype. Typically, the service parameters of SHS products are 3 - 5 times higher. For the SHS powders, their composition (chemical, phase, and granulometric) and the properties of end products (produced from powders) are compared with those of their furnace analogs.

Of great importance here are the technological factors, first of all, the economic factors. As a rule, a comparative analysis is attended by serious difficulties due to a lack of technological details (know-how).

Using the Russian experience in SHS production, the production cost for some SHS powders was calculated with due regard for the Western conditions (the cost of raw materials, labor cost, etc.). It was found to be lower than the world market prices of these products; this confirmed high economic efficiency of SHS technologies. The ratio of the cost of SHS products to the world prices is 0.13 - 0.25 for the elemental powders (AlN, Si_3N_4); 0.25 - 0.35 for aluminothermic powders (Cr_3C_2); and 0.4 - 0.45 for magnesiothermic powders (TiB_2, BN).

None of the SHS productions can be set up without a thorough economic design.

ELECTRIC PROSPECTING FOR OIL AND GAS FIELDS USING POWERFUL MHD GENERATORS

J. A. Besruk, B. P. Zhukov, E. P. Velikhov, Y. P. Babakov

101000 Moskow, 22 Chernyshevsky St., Vneshgeofirma

410710 Saratov, 70 Lenin avenue, IVNIIGG

Resume

Achievements of the seismic survey for reconnaissance of anticlinal traps of oil and gas has depleted practically the reconnaissance reserves. Future prospects of oil and gas field discovery are connected now with nonanticlinal traps having high electric conductivity because of saturation by mineral water, owing to this there is an electric conductive contrast agains a background of adjacent layers. At the same time seismic survey badly control nonanticlinal traps by surface layer interfaces.

Problems

More deep exploration methods are required for nonanticlinal traps because of occurrence conditions. Highly reliable geophysical reconnaissance is needed owing to high prices of occurrence opening by boring. The complex method of the reconnaissance is imminent which includes a seismic survey (for exact depth cartograghy) and highly productive area electric prospecting (for traps detection and estimation of their collector properties and exact position).

"3 C" Method with MHD generator

High productive electric prospecting according to "3 C"-MHD technology is given shematically in Fig. 1. Electromagnetic field on the sounding area is generated by earthed cable-dipole with length of 10 km, the dipole for 5...10 s is switched to the source-mobile MHD generator with power of 16 MW, generating current up to 10 kA (MHD generator

H. Krause (ed.),
***Conversion Concepts for Commercial Applications and Disposal Technologies of Energetic Systems*, 85–90.**

"start"). Sounding field at the distance up to 30...50 km from dipole is recorded by multichannel radiotelemetrical recording system (100 and more channels). The system elements are located according to the grid of 2 x 0,5 km. Capacity of the complex with MHD generator is up to 50 km/start (100 km^2 / start) at the sounding depth up to 8...10 km.

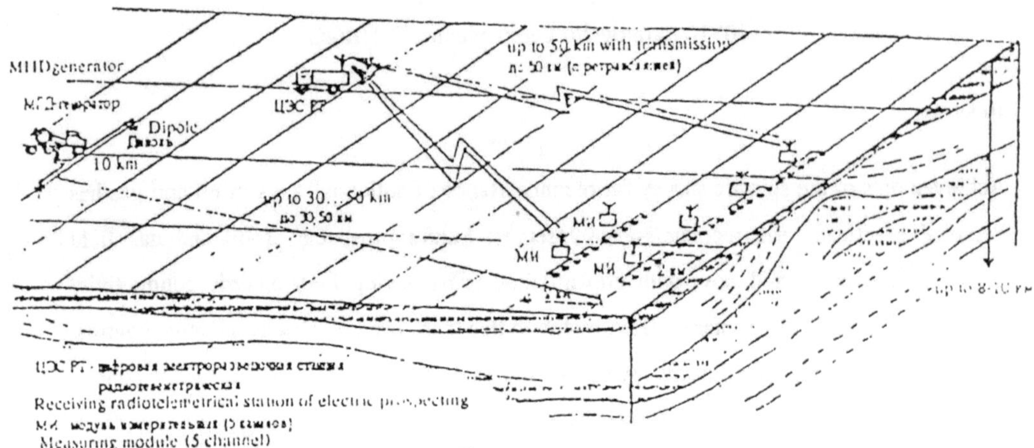

Fig. 1.

Using the electric prospecting with the powerful source in the geological prospecting complex raise the coefficient of successful bore with 0,46 up to 0,67.

At exploration the area up to 5000 km^2 (up to 50 starts) MHD generator and dipole are not moved that permits to use small-scale radiotelemetric apparatus without harm to agriculture fields and nature, with ecology safety.

Fig. 2. and 3. show the results of the "3C"-MHD technology application in the Sub-Caspian. Layers-collectors with electric conductivity have been detected at the depth of 5...8 km and the object of the "reef" type has been discovered in them. Using the seismic

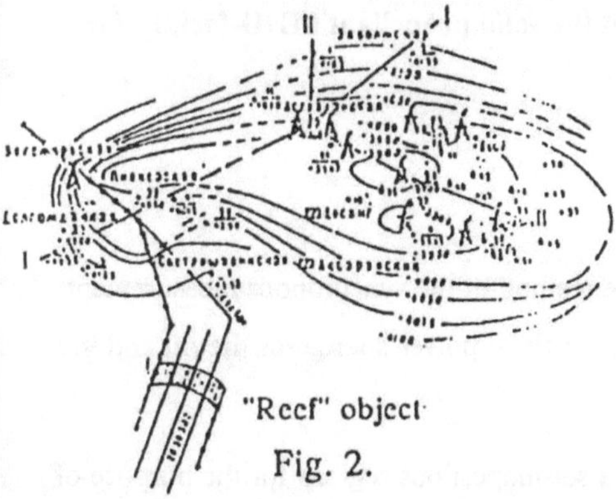

Fig. 2.

Geological profile according to line I—I

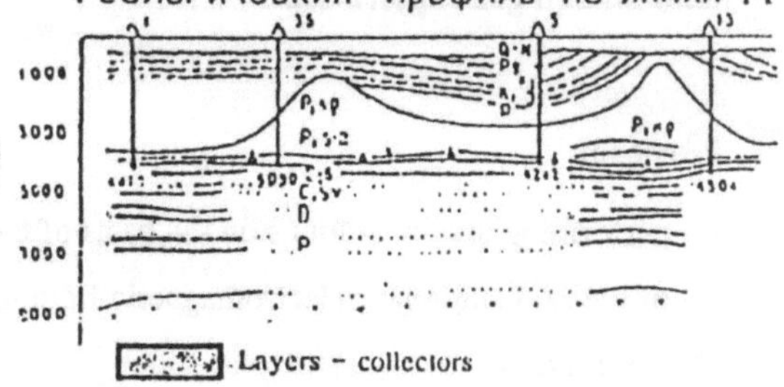

Fig. 3.

survey (exact depth definition) the object has been bored. The industrial oil flow has been obtained at the depth of 5,5 km. This example shows the greater depth of the "3C"-MHD technology owing to powerful sounding current pulse and high discovery reliability of complex nonanticlinal objects which are perspective for reconnaissance of oil and gas fields. These objects are discovered reliably by the complex: electric prospecting (collector discovery, Fig. 4.) and seismic survey (deep structures, Fig. 5.).

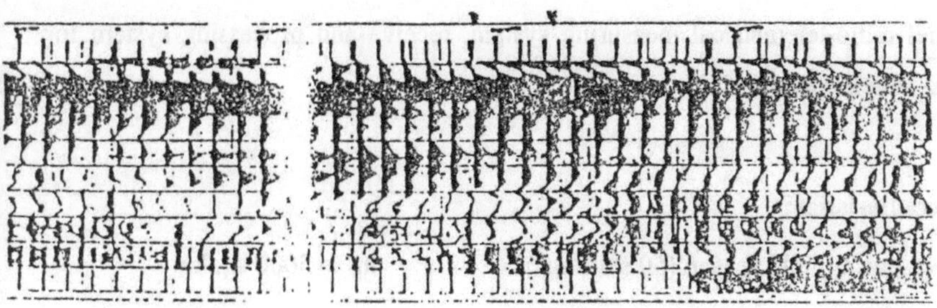

Electric conductivity of "reef" type object according to electric prospecting data.

Fig. 4.

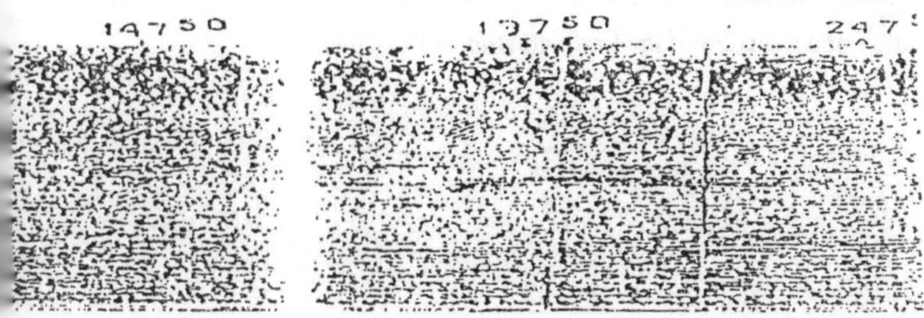

"Reef" object structure according to seismic survey data.

Fig. 5.

Geophisical Complex "Sojuz" on the basis of the solid propellant MHD-facility for oil land and gas exploration

Purpose:

- performing area operations on the basis of the method of the synchronous measurements of the non stationary electromagnetic field using fixed power source for the oil and gas exploration;
- regular control of a geological parameters in a seismoperilous regions for the purpose of prediction of earthquake;
- investigation of a depth structure of the Earth crust;

Structure of the complex:

- MHD-facility (SP MHD-generator and transforme and control unit);
- Electrical dipole with the real-unreal device;
- Multichannel radio-telemetrical measuring system, receive and processing system for geophysical information.

Technical characteristics:

Prospecting area during field season (50 startings)	up to 5000 sq. km
Depth of a cartography of gas and oil deposits	up to 8 km

Depth of investigation of a structure of the Earth crustup to 15 - 20 km

MHD-facility:

power in the load	up to 16 MW
current in the load	4 - 7 kA
duration of the current pulse	10 s

Electrical dipole:

length	up to 10 km
active resistance	0.3 - 1.0 Ohm

Radio-telemetrical measuring systems:

measuring parameters	components of the electromagnetic field
number of channels	100 and more
frequency range	0.001 - 100 Hz
distance of date transmission	up to 50 km
data processing	using IBM PC
Construction of the complex	mobile.

The technology of area investigation offers high geological efficiency which is achieved due to high characteristics of the complex components, application features special techniques for processing algorithmus and programs, as well as analysis of high-resolution differential curves of the field formation process.

Scheme of geophysical "Sojuz" SP MHD-complex.

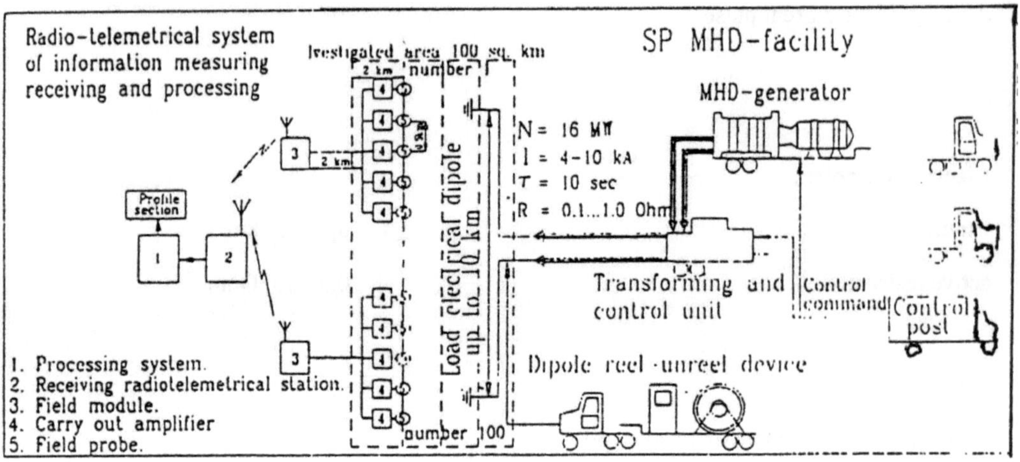

ADVANCED DISPOSAL AND RECYCLING TECHNOLOGIES OF ENERGETIC MATERIALS

H.H. Krause
Fraunhofer-Institut für Chemische Technologie (ICT), P.O.Box 12 40, 76327 Pfinztal, FRG

Abstract

Due to the reduction of armament between the East and West, we were met by new objectives: the disposal of energetic materials.
Environmentally compatible disposal methods available for the different energetic materials are highly requested. From the technical point of view recycling processes are generely available, but the market for recycled explosives is very limited. The possibilities of reprocessing propellants and explosives have to be reexamined.
The most advanced disposal technology is incineration. Rotary kiln as well as fluidized bed installations were successfully tested. The main problem on incineration remains the reduction of pollutant emission below the legal limits.
Chemical decomposition of explosive materials followed by biological degradation of the reaction mixture offers an attractive alternative to incineration processes. The initial chemical step delivers a biodegradable reaction mixture. The combined chemical/biological process was successfully tested with NC based gun propellants. Application tests with other explosives are in progress.
For toxic and hazardous components of pyrotechnical ammunition supercritical water oxidation was applied. Especially chloro organic compounds can be destroyed in an environmental friendly way by SCWO-technology.

1. Introduction

The reprocessing and/or disposal processes are tied up with the composition and thus with the chemical properties of the different materials involved.
Waste disposal legislation demands the precedence of reprocessing before pure disposal methods. Therefore processes where a reuse of energetic materials can be carried out in an economic and ecologically viable manner should be preferred.
The disposal and recycling processes for explosives must be

- safe,
- technically feasible,
- environmentally safe (ground, water, air, noise factor) and
- economic in their operation.

H. Krause (ed.),
Conversion Concepts for Commercial Applications and Disposal Technologies of Energetic Systems, 91–97.

2. Reprocessing of Explosives

2.1 Recycling of TNT

Technologies designed to melt out TNT are well known and available. After cleaning and preparation, e.g. flaking, the TNT can be reused by the explosives industry.
The TNT thus obtained will most probably be used in explosives for commercial application. For military reuse it would not meet with quality demands of the MIL Standard without passing through further purification steps. Through further processing and/or purification steps (washing with solvents, recrystallization, treatment with adsorbent agents etc) higher qualities are concievable. The technologies for such processes are available [1] but were mostly not realized.

2.2 Reprocessing of TNT

There are a number of proposals on the use of TNT (as well as dinitrotoluene DNT) as a chemical raw material. Examples for this are:

- hydrogenation of TNT to triaminobenzene (TAT)
- hydrogenation of DNT to diaminobenzene (DAT)
- DAT and TAT phosgenation to produce isocyanates.

In the case of mass industrial products such as isocyanates, it must be remarked that the quantity of TNT available is relatively small in comparison with the commercial isocyanate production. The chemical transformation of TNT to TAT was successfully demonstrated on laboratory scale, but an industrial market for TAT doesn't really exist.

2.3 Reprocessing of nitrocellulose to make lacquer NC

Nitrocellulose gun propellants (single-base) can be reprocessed to lacquer NC. For this purpose, the propellant has to be grinded before dissolving it in acetone, after which the NC is precipitated out and separated. The decisive process step is the boiling of the NC in ammonia to reduce its nitrogen content to below 12.6 %. In this step as well as in the subsequent acid boiling under pressure, the decrease in molecular mass of the NC must be controlled in such a way that viscosities are obtained in accordance with Type Production Guidelines for lacquer NCs (German Industrial Standard DIN 53 179).
This process (according to laboratory results) supplies lacquer NC containing impurities as diphenylamine and decomposition products.
The process fulfills the reprocessing concept: its outlay is similar to that for the manufacturing of lacquer NC from primary materials, although it results in a product of poorer quality. In addition to this we must consider the fact that the market for nitrocellulose of this type has been in a recess for years now, and that it is only possible to market qualitatively high-value products.

3. Disposal of Explosives

The disposal processes are at the end of the reuse cascade:

- product
- component
- raw material
- energy

Here, if at all, only the energy content of the substances is made use of.
In the case of the pure disposal processes, therefore, the elimination of the explosive character in a manner least harmful to the environment should be the prime objective.

3.1 Chemical/biological disposal of propellants

This process is suitable for all explosives containing nitrocellulose as a major component.
The chemical/biological disposal of NC based propellants consits of three steps:

- alkaline hydrolysis
- anoxic biological treatment (denitrification)
- aerobic biological treatment.

When using larger propellant grain sizes or rocket motors, milling or grinding processes should first be carried out in order to operate the process continously on the basis of pumpable slurries. An exact control of the hydrolytic process is of great importance, as the substances are here prepared for the subsequent biological steps. The aim is to obtain the best possible biodegradability of the reaction mixture.
Pressure hydrolysis (< 2.0 MPa) at temperatures between 130 °C and 150 °C is particularly well suited. This method produces smaller decomposition products on a molecular level, i.e. species which are biologically more favorable for further reduction. The main products in the aqueous reaction mixture are nitrite, nitrate and small hydrocarbon acids (like formic, oxalic, bernstein or glutaric acid).
As gaseous by-products CO_2, N_2, N_2O and NO_x were formed during the hydrolysis. The nitrogen the propellant contained is converted to NO_2^- and NO_3^- (40 - 50 %) and to N_2 and N_2O (30 - 40%). NO_x formation doesn't play a role with values below 0.1 % N-conversion.
Problems with this process may be caused by additives of propellants such as diphenylamine (DPA), DNT, Centralite or heavy metal compounds. These substances should not be transported via the process water or the sludge in the environment. They must be already destroyed in the hydrolytic stage or be transformed to harmless products in the subsequent biological purification. In the case of propellants which contain heavy metals as burning modifiers (cupper and lead compounds), the heavy metals can be easily removed after the hydrolysis stage.

Should the propellant contain chlorinated resins, the mixture for hydrolysis should additionally be subjected to a SCWO (supercritical water oxidation) treatment. With this method, destruction efficiencies of over 99.98 % are obtained for halogene containing hydrocarbons.
The excess biological sludge (< 10 % of the input mass) must be disposed of. The possibilities of using the sludge as soil improver are rather good, if the raw material didn't contain heavy metals.
The chemical/biological process is suitable for a large number of NC propellants to be disposed of. Care must be taken, however, as regards the propellants which contain critical components.
The process has advantages over incineration procedures as regards the granting of legal permission (acceptance by the general public).

3.2 Incineration

3.2.1 Problems in the incineration of explosives

The chemical nature of explosives implies that, on burning under normal pressure, large quantities of NO_x as well as soot can be produced. Contrary to other incendiary materials containing nitrogen, explosives already produce considerable quantities of NO_x on their own thermal decomposition.
The upper limit of possible NO_x emission depends on the pyrolytic behavior of the explosive concerned. Every second group of $O\text{-}NO_2$ produces a NO_x under conditions of pyrolysis. On the admixture of air, according to the burning conditions, the formation of NO_x is reduced by a factor of 5-20. The NO_x emissions on burning with additional air supply are typically at 5 000 to 10,000 ppm.
The most favorable values are attained at lambda values of 0.6 to 0.8.

3.2.2 Incineration concepts for explosives

Different incineration systems are already in use for the disposal of propellants and explosives:

- Rotary kiln (Burlington AEC-Plant) [2]
- Fluidized bed incinerator (ARRADCOM) [4]
- APE 1236 Deactivation furnace (Toole army Depot) [3]

The APE 1236 furnace, a so-called armored rotary kiln, is suitable for the disposal of infantry ammunition without previous dismantling up to 20 mm (caliber).

Table 1: Limit values of the 17th German Federal Immission Protection Regulation (17.BimSchV)

	24 hours (mg/m^3)	hour (mg/m^3)
CO	50	100 (1 hr)
Dust	10	30
C total	10	20
HCl	10	60
HF	1	4
SO_2	50	200
NO_x (NO_2)	200	400
Cd, Tl		0.05 mg/m^3
Hg		0.05
Sb,As,Pb,Cr,Co,Cu,Mn,Ni,V,Sn:	0.5	
Sum of dioxins and furanes (for 11 % O_2)	0.1 ng/m^3	

Both rotating cylinder furnaces as well as fluidized bed incinerators are suitable for disposal of explosives.

Rotating cylinder furnaces are limited as regards control of the combustion process, although they offer the advantage of being omnivorous. At raw gas concentrations of 10,000 ppm NO_x, high-outlay denitrating processes turnover rates above 95 % must be applied to meet legal requirements.

The special armored rotating furnace is a method of choice where whole ammunition (up to 20 mm) is to be disposed of without previous dismantling. However, this process in not suitable where ammunition containing mercury is involved.

Fluidized bed incinerators possess a higher potential in the specific control of the incineration process (stepwise incineration).

The application of NO_x-reducing additives in combination with a stepwise incinerating process ought to make NO_x emissions possible in a range of 500 to 1 000 ppm, as laboratory tests at the ICT have shown [5]. The combination of this technique with a fluidized bed combustion system should therefore be pursued further.

3.3 Disposal of critical components by SCWO

SCWO stands for supercritical water oxidation. The critical point of water is at 372.2°C [701.96°F] and 21.8 MPa.

On oxidation in supercritical water:

- over 99.9 % of organic chlorine compounds are destroyed and converted to harmless chloride,
- up to 99.9 % of any existing dioxin is destroyed,
- anorganic substances are separated off,
- over 99 % of organic carbon is oxidized to CO_2.

In USA and Germany there are starting research programs to solve the basic engineering problems of SCWO-technology on a pilot plant level.

- corrosion
- salt formation
- continuous feed of slurries
- effluent solid and separation

Thus, this process is particularly suitable for the disposal of very critical substances such as oils containing PCB as additive, chlorine containing binders or toxic and hazardous materials found in old ammunition.

4. Conclusion

Waste disposal legislation demands the precedence of reprocessing before pure disposal methods. Therefore processes where a reuse can be carried out in an economic and ecologically viable manner should be preferred.

- Recycling

The direct product recycling of energetic materials is technically realized, but in praxis the market for regenerated explosives is very limited. Only a considerable amount of TNT is reused in explosives for commercial application.

- Reprocessing to chemical raw materials

The chemical feasability of the NC-conversion to laquer NC and the TNT Hydrogenation to TAT was technically realized.

Nevertheless the marketing (cost/benefit ratio) of the products is the limiting factor.

Generally recycling and conversion of energetic materials play in praxis not an important role, whereas advanced disposal techniques in addition to incineration were used to destroy energetic materials.

- Incineration

The praxis of open burning or open detonation is today obsolet for environmental reasons. Rotary kiln as well as fluidized bed incinerators are available to destroy energetic materials in an environmentally friendly way. The high NO_x emission of energetic materials requires an additional DENOX stage to fulfill the limiting values of legal regulations. Incineration is the state of the art disposal method for energetic materials.

- chemical/biological disposal

The chemical treatment of energetic materials to generate a biodegradable reaction mixture is an alternative to incineration for NC based materials. The application of this combined technique should be expanded to other energetic materials like TNT, RDX, HMX etc.

- Supercritical water oxidation (SCWO)

This advanced new disposal technology is ideal to destroy with highest destruction efficiency very toxic and hazardous materials. This upcomming new technique need further engineering development to become an standard disposal method.

REFERENCES

[1] H. Herman
Die Vernichtung nicht mehr benötigter Sprengstoffe aus Herstellung und Konversion - am Beispiel TNT, 22nd International Annual Conference of ICT, 1991

[2] Worley, R.D.
Disposal of Waste or Excess High Explosives, Mason & Hanger Report MHSMP-75-14, April 1975

[3] Anderson, R.G.
APE 1236 Deactivation Furnace Upgrade to meet RCRA-Requirements Contaminated Soil 90, Vol. II, 1435 - 1438

[4] Scola, R.; Santos, J.S.
Fluidized Bed Incinerator for Disposal of Propellants and Explosivses, NTIS HC A07/MF A01 CSCL 19/1

[5] Gröbel V., Eisenreich N., Krause H., Weiser V.
Feuerungstechnische Untersuchungen des Abbrandverhaltens und der Schadstoff-emissionen von Treib- und Explosivstoffen, ICT-Bericht 9/91

DEVELOPMENT OF SOLID PROPELLANTS FOR EFFECTIVE FIRE-EXTINGUISHING SYSTEMS

Denisjuk A. P., Zhukov B. P., Alferov V. G., Shepelev Yu. G.,
Baloyan B. M., Rusin D. L., Sedov S. A.

INTRODUCTION

Uncontrolled combustions and explosions result in great losses for the mankind. One of the reasons for this is the low effetiveness of the facilities used to fight fire.

One of the developments of new fire-preventing techniques is the elaborating of fire-suppressing systems with great effectiveness, which may be obtained by using special solid propellants.

The working body formed as a result of the combustion of the propellant, depending on the type of a system, can function as follows: it either inhibits chain reactions in a flame by means of aerosol, or displaces a pre-made fire-suppressing agent out of a can and then spray it through out in air, or dilutes atmosphere in the site of origin for fire with non-explosive and non-combustible products of burning (CO_2, N_2 H_2O).

The research work is carried out on a large scale in different scientific centers in Russia to evolve powders and pyrotechnics to be used in a number of fire-extinguishing systems [1-4].

This paper is devoted to the development of aerosol compositions and propellants with minimal content of combustible gases in the products, which are manufactured according to the extrusion technology.

H. Krause (ed.),
Conversion Concepts for Commercial Applications and Disposal Technologies of Energetic Systems, 99–112.

AEROSOL FORMING PROPELLANTS

While elaborating such propellants, a wide experience was used for minimizing the muzzle flash occurring during combustion of the propellants with the help of potassium-containinig components (5). According to (2), the fire-extinguishing efficiency of the compositions that form potassium-containing aerosol (1,6) 3-12 times surpasses on burning the widely-used Freons, the application of which is practically prohibited due to their hazardous effect on the Earth's ozone layer.

Thermodynamics of the compositions

The thermodynamic characteristics were calculated for a limited number of the compositions aiming the development of the products possessing the acceptable physicomechanical, rheological and ballistic properties.

The compositions developed were based on the following binders: nitrocellulose (NC), triacetin, nitroglycerin (NG), dibutyl phthalate. Potassium nitrate was used both as a flash reducer and oxidizer. Its content must be considerable (55 % and more) to provide high fire-extinguishing ability. High-dispersed soot was used as a fuel. The ratio plasticizer-polymer (P/NC) was 1.0-1.2, the total amount of fillers were up to 80 %.

The compositions containing 50 % KNO_3 based on NG even with the extra plasticizer added have the combustion temperature too high for fire-extinguishing purposes (1600-2000 K) and do not contain any solid phase in the combustion products. Virtually, all the potassium is converted into the gaseous KOH.

Table 1

NC-triacetin-based propellants burning temperature and composition of the combustion products (P/NC=1.2)

filler content and composition				Condensed products, %		Gaseous KOH content, %
%	η	α	T_c, K	K_2CO_3	Carbon	
50	7,5	0,51	1032	30,1	0,9	traces
60	7,5	0,58	1200	36,0	no	0,04
70	7,5	0,66	1423	36,9	no	1,6
80	7,5	0,78	1541	32,5	no	5,6
75	5,0	0,59	1270	41,9	no	0,1
75	7,5	0,72	1486	34,9	no	3,4
75	10,0	0,80	1551	32,5	no	6,8

Combustion temperature of NC-tryacetin-based composition increases when the content of a filler and ratio of KNO_3 / soot (designated as η) increases (see Table 1). This growth corresponds to the increase of the stoichiometric ratio α. The solid phase mainly consists of K_2CO_3 and solid carbon, which begins disappearing at the temperature higher than 1100 K, simultaneously KOH appears in gaseous products.

The maximum content of condensed K_2CO_3 and the practically complete absence of KOH at lowest temperature of combustion is observed when the ratio η is about 5. However, considerable (12.5 %) amount of high-dispersed soot is undesirable due to difficulties with the production of the propellant possessing the satisfactory rheological characteristics. The analogous results were obtained for the propellants based on NC-dibutyl phthalate and other binders.

On the basis of analyzing the data of Table 1, the model propellant was developed based on NC-triacetin binder at 75 % filler and η in the range 5-7.5. The propellant yields a big amount K_2CO_3 and small KOH in the products, its adiabatic temperature is below 1500 K.

First of all, it is necessary to study the model's burning characteristics and its fire-extinguishing ability (FEA).

COMBUSTION PROCESS CHARACTERISTICS

Influences of η on the burning rate

Soot with specific surface 75 m^2/g and highly-dispersed potassium nitrate (particle size approx. 10 mcm) were used in the propellants investigated.

The value of η effects greatly the burning rate and the stability of the combustion of the propellant under atmospheric pressure (P). The propellants with η=5 and η=6.5 under atmospheric pressure have the burning rates 7.0 mm/s and 3.8 mm/s, respectively (Table 2). The propellants with η=7.5-10 being put into charges diameter 6 mm burn unstable and usually extinguish.

Under higher pressure (0.3-3.0 MPa), the decrease of η from η=10 to η=5 causes the increase in the burning rate along with the reduction of the dependence of the burning rate on the pressure (the ν number in the expression $U=BP^{\nu}$).

Thus, the ratio KNO_3 / soot is the factor which determines the value of the burning rate and the stability of burning under atmospheric pressure.

Influences of soot dispersity on the burning rate

The combustion of the propellants with η=7.5 was studied. The potassium nitrate + soot mixture amount was varied from 60 to 80 % (P/NC=1.2.). Soots with S_s 250, 100 and 75 m^2 / g were used.

Samples of a diameter of 6 mm and containing a mixture of 60-75 % KNO_3 with soot of different dispersity fail to burn under atmospheric pressure. The combustion is observed only when 80 % of mixture KNO_3 / soot is present in the propellant. However, such composition are technologically ineffective. If ignited at the side surface, the plates made of those propellants burn completely.

Table 2

Burning rate and FEA for the different KNO_3 / soot ratio propellants

KNO_3 / soot ratio	U, mm/s under P, MPa					ν within interval 0.3 - 3 MPa	Extinction time τ, s (m=0.8 g)
	0,1	0,3	0,5	1,0	3,0		
5,0	7,0	9,0	9,0	9,0	9,0	0	57
6,5	3,8	6,5	7,0	7,8	9,5	0,16	31
7,5	-	6,5	7,2	8,2	10,3	0,20	27
10,0	-	4,3	5,2	6,9	10,5	0,40	-

Addition of 10 % of octogen or potassium perchlorate to the composition of the propellant containing highly-dispersed (250 m^2/g) soot does not result in a steady combustion under atmospheric pressure. The steady-state combustion might only be achieved when a catalyst is used. The sample with 5 % catalyst burns at a rate of 3.9 mm/s under the pressure of one atmosphere. The value of ν within the pressure interval 0.25-3 MPa is close to zero. It is necessary to note, that the catalyzed sample burns faster than that without the catalyst only at a pressure below 0.3 Mpa. Above this pressure the catalyst has no influence on the burning rate.

The further investigations of the effect of soot dispersity on the burning rate were carried out for the propellant containing 5 % of the catalyst.

From the Table 3 it is seen that the low-dispersed soot samples fail to burn under atmospheric pressure, though they burn at the rate as high as 6 mm/s for pressures above 0.3 MPa.

Within the pressure interval 0.3-3.0 MPa the burning rate increases as the specific surface of the soot increases. Thus, under pressure of 0.3 MPa the burning rate of the samples with soot S=250 m^2/g equals 14 mm/s, whereas in case of soot S_s=75 m^2/g it equals 6,5 mm/s. The value ν for these samples equals 0 and 0.2, accordingly.

The experimental combustion temperature was determined for the propellant samples (catalyzed and noncatalyzed) containing soot with $S_s = 75\ m^2/g$ (Table 4).

Table 3

Catalyzed propellant (5 % catalyst) burning rate dependence upon soot dispersity

S_s, soot, m^2/g	Burning rate, mm /s under pressure, MPa					ν (0.3-3) MPa
	0,1	0,3	0,5	1,0	3,0	
250	3,9	14,0	14,0	14,0	14,0	0
100	fail to burn	8,2	8,9	9,9	11,8	0,15
75	fail to burn	6,5	7,2	8,2	10,3	0,20

Additionally, these samples contain 1.7 % technological constituent, which simultaneously assures a steady-state combustion under atmospheric pressure and without a catalyst. The influence of this additive is studied in details below.

Table 4

Propellant combustion temperature

Pressure, MPa	Noncatalyzed		Catalyzed		
	$T_{calc.}$,K	$T_{exp.}$,K	$T_{calc.}$,K	$T_{exp.}$,K	$Z = U_k/U_o$
0,1	1375	1515 1466	1343	1304 1327	1,47
1,1	1478	1627 1663	1417	1530 1543	0,94
2,1	1493	1642 1627	1424	1627 1644	0,98

For a noncatalyzed sample the measured combustion temperature (T_c) increases as the pressure rises from one atmosphere to approx. 1 MPa and then becomes constant. However, of the calculated value of T_c upon depends on pressure, the calculated temperature is lower than what was observed in the whole pressure range: under atmospheric pressure approx. by 110 K, under 2 MPa by 140 K.

In the pressure interval considered, calculated and experimental combustion temperatures of catalyzed composition (Table 4) were found to be below those for the sample without a catalyst. Under atmospheric pressure, when a catalyst considerably accelerates combustion of a propellant (Z=1.5), the values of experimental and calculated combustion temperatures differ slightly. Under a pressures above 0.5 MPa, when the catalyst's influence is diminished experimental combustion temperature becomes higher than calculated. This may be explained by the fact that in combustion the thermodynamic equilibrium is not reached and thus there is carbon in the combustion products (in the form of either original soot or yielded by propellant constituents decomposition), which fails to react completely in the combustion wave in accordance with the equations

$$C + H_2O \leftrightarrow CO + H_2\ \Delta = +\ 132.7\ \text{kJ/mol.}$$

$$C + CO_2 \leftrightarrow 2\ CO \qquad \Delta = +\ 172.6\ \text{kJ/mol.}$$

which proceed to absorb heat. It is confirmed by the calculations made for the constant carbon content. Carbon in amounts of 3.6 % is present in products. The calculated temperature (1632 K) is practically the same as obtained from experiments. The following variations of combustion products content take place: H_2O increases from 1.2 to 3.0 mole/kg, CO_2 does from 1.2 to 2.4 mole/kg, H_2 decreases from 5.5 to 3.6 mole/kg and CO diminishes from 13.1 to 9.1 mole/kg. From here, it is seen that there is actually a contribution of both reactions, while the first of them has a greater influence than the second one.

FIRE-EXTINGUISHING ABILITY OF COMPOSITIONS

The tests of extinguishing the flames of combustible liquids -(isopropanol, etc.) were made in a 69 liters box, in which a burner and a fan for averaging the aerosol concentration were placed. The fire-extinguishing ability (FEA) for the patterns was obtained from the dependence of the burner extinction time upon the propellant sample weight and was characterized by the value of $\Upsilon = m^* /V$, where m^* is the propellant sample weight at which there is an abrupt bend on the $\lg\tau=f(m)$ dependence, V is the box volume.

One of the propellants was tested to estimate its FEA, when used to suppress the burning of various substances. A forming aerosol proves to be able to suppress the burning of all liquids used, which differ by combustion heats, boiling temperatures and other properties. For an extinction to occur, different quantity of a propellant is required per 1 m^3 of volume: for isopropanol - 8.9 g, for acetone - 9.9 g, for gasoline - 14 g, for kerosene - 15.4 g, for tetradecane - 16.8 g, for ethylene glycol - 24.4 g, for diethylene glycol - 27.3 g.

As it was shown by electron-microscopic investigations, the particles released by aerosol after burning have an irregular shape (form factor 0.75) and a mean size of 2.2 mcm. The main part of the sample (approx. 75 %) consists of the particles with sizes ranging 1.7-3.4 mcm. By spectroscopic analysis it was stated that the particles composition is represented mainly by $KHCO_3$.

All further experiments of FEA determination were carried out with isopropanol. Influence of the ratio KNO_3 / soot (ranging 5-10) as well as soot specific surface on the FEA was studied. It is seen from Table 2 that the propellants with the higher values of η possess better extinguishing characteristics. The least extinction time was found to be provided when the ratio η was 7.5 and 5 % of a catalyst added. In general, the ratio KNO_3 / soot ranging from 6.5 to 7.5 is the optimal one.

The fire-extinguishing ability for the samples containing soots S_s = 75 and 100 m^2 /g differs slightly, but it is still lower than that for the sample with the high-dispersed soot (S_s = 250 m^2 /g.)

Thus, the best results (FEA and combustion stability) were obtained , when high-dispersed soot (S_s = 250 m^2 /g) and catalyst were used and the ratio KNO_3 / soot was 7.5.

OPTIMIZING THE COMPOSITIONS

The investigations of the propellants have shown that they have a poor adaptability to manufacture, especially those containing the highly-dispersed soot. For these reasons, the content of the filler in the propellant was reduced to 72.8 %, the soot S_s = 75 m^2 /g was used, fluorine-containing additive and sodium stearate were added.

The fluorine-containing additive, even in a small amount, greatly affects the combustion of a propellant under pressures close to atmospheric (table 5). Without this additive, propellant samples of diameter 6.5 mm burn steadily only beginning from 0.3 MPa. The addition of as little as 0.2 % additive brings about the steady-state combustion under atmospheric pressure at a rate as high as 5.6 mm/s. The dependence of the burning rate upon additive content is the extremal one, having a maximum at a content of 1 %. The fluorocontaining additive is also responsible for significant improvement of physicomechanical (Table 5) and technological characteristics of the samples. In order to achieve an acceptable level of physicomechanical and rheological properties, the optimal content of additive was found to be 1.5 - 1.7 %.

Table 5

PROPELLANT BURNING RATE DEPENDENCE ON FLUOROCONTAINING ADDITIVE CONTENT

Additive content %	Burning rate, mm/s, under pressure (MPa)			Physicomechanical properties		ν (within the interval, MPa)
	0,1	0,2	0,3	σ_t^{20}, MPa	ε_t^{20}, %	
0	-	-	5,0	3,5	13	0,21 (0,3-1,0) 0 (1,0-3,0)
0,2	5,6	-	6,3	3,6	14	0,25 (0,3-3,0)
0,5	5,7	6,5	7,0	4,2	24	0,17 (0,1-3,0)
1,0	5,8	6,8	7,4	5,4	24	0,19 (0,1-3,0)
1,7	5,4	6,2	6,6	7,8	33	0,29 (0,1-0,2) 0,18 (0,2-3,0)
2,1	5,4	6,3	6,7	9,7	48	0,29 (0,1-0,2 0,18 (0,2-3,0)

Such propellant was studied to investigate the influence of KNO_3 dispersity on burning rate and fire-extinguishing ability.

The propellants based on moderately-dispersed (160-315 mcm) and high-dispersed (approx. 10mcm) potassium nitrate were produced. The electron-microscopic investigations of KNO_3 dispersity in a formula of the final propellant were carried out to show that there was a substantial grinding of large-sized particles (original size 160-315 mcm) in the process of propellant preparation. After having been rolled with a clearance of 2 mm mean particles size turned out to be 15 mcm, whereas at clearance of 0.5 mm it was as little as 7 mcm. The above-said roll clearances provided, high-dispersed KNO_3 (approx. 10 mcm) practically is not divided. Because of the considerable grinding a large-sized KNO_3, the samples based on it differ relatively slightly in burning rate from those, based on high-dispersed KNO_3. Though, there is a correlation between the burning rate and KNO_3 particles sizes in the final propellant. If rolled at a clearance of 2 mm, a sample based on large-sized KNO_3 burns nearly by 30 % slower than the sample rolled at 0.5 mm, and by

60 % slower than the sample based on high-dispersed KNO_3. Besides it is stated, that the higher the dispersity of KNO_3 is in a propellant, the better is its FEA.

Thus, FEA of the samples containing high-dispersed KNO_3 is higher than that of low-dispersed.
Except for the compositions on the basis of NC-triacetin blend, the properties of the propellants based on more thermoresistant binders NC-dibutyl phthalate and polyvinyl butyral (PVB)-dibutyl phthalate were investigated. As in case of NC-triacetin propellants, these compositions demonstrated the same character of effects of potassium nitrate and soot dispersity, their ratio and total amount on the burning rate, FEA, physicomechanical characteristics, etc.

In general, the investigations showed the possibility of development of high effective aerosolforming compositions of a double base propellants type, which would contain a high content of fillers, burn steadily under low pressure (beginning from atmospheric), possess a light dependence of the burning rate upon pressure and demonstrate acceptable technological properties. The whole complex of the characteristics mentioned is provided by choosing the optimal values of P/NC, KNO_3 and soot content and dispersity, using specially selected combustion catalyst and complexible modifier.

"INERT" GASES GENERATORS PROPELLANTS

One of the principal requirements these propellants meet is a limited (no more than 10 %) content of CO and H_2 in the products, because of the ability of these gases to form combustible and explosive mixtures with air. There is a 4-7 times larger amount of the pointed gases in standard propellants used today. It apparently seems, that those requirements cannot be satisfied without a significant amount of oxidizer being added to the propellant composition Potassium nitrate again was used.

In the beginning, high-energetic mixtures of NC+NG were used for reducing the amount of oxidizer in the composition and obtaining a high yield of gases without producing a large amount of solid phase in the burnt products. The oxidizer excess coefficient α was 1.01.

It is necessary to note the principal results of the calculations. There are 0.5 % H_2, 5.2 % CO and 3.9 % O_2 in combustion chamber under pressure of 2 MPa, whereas under 7 MPa the content of these products decreases to 0.4 %, 4.4 % and 3.3 %, respectively. In the outlet gage of the nozzle there is a great decreasing of the total amount of the pointed gases. Their content directly corresponds to the coefficient of expansion. An excess of oxygen with α=1.1 does not eliminate the mentioned combustible gases, when as reduction of α to 0.96 causes an abrupt vise of the CO content (up to 15 %).

At α=l, the reason of the formation of CO and H_2 in combustion products is the dissociation of water and CO_2 because of a high combustion temperature. To prepare a propellant with lower combustion temperature diethylene glycol dinitrate (DEGON) was taken instead of NG.

First of all, a number of experiments were carried out with a model propellant containing 55 % NC and 45 % DEGON. The combustion temperature (T_c) of this base in the chamber is 2270 K, at the nozzle end 1130 K, at Pc/Pe = 4/0.1, where Pc is the chamber pressure and Pe is the pressure at the nozzle end. If potassium nitrate (10-50 %) is added to the composition (α=l at 49 % of KNO_3), the value of Tc increases up to 2419 K in the chamber and 1626 K at the nozzle end, whereas the total content of CO and H_2 falls down to become 1.2% in the chamber and 0 % at the nozzle end.

Depending on its content, the oxidizer has a different influence upon the binder burning rate: 30 % KNO_3 enlarges the burning rate by 15-25 % and 50 % - brings it down by 40-50 %. For all the samples concerned, the value of ν ranges 0-7-0.8 under pressure 2-10 MPa. Under relatively low pressure (approx. 2 MPa) the combustion of KNO_3-containing samples is pulsatory and only partial thermolysis of potassium nitrate occurs, what means that the propellant does not burn completely. The combustion process for such a composition is highly affected by copper-lead phthalate (CLP) combined with soot or fluorocontaining modifier.

With the total amount of additives nearly 3-5 % the growth of the burning rate in near 60-70 %, whereas ν varies just a little.

One of the reasons why a catalyst has no significant influence on the value of ν of the propellant is a relatively low effectiveness of its action upon the binder burning rate and ν. In addition, the binder used does not provide acceptable technological characteristics. Therefore, further experiments dealt with the propellant with the enlarged ratio P/NC=1.3,at a=l (55 % KNO_3).

Because soot greatly diminishes α, catalysts with fluorocontaining additive were used to control the burning rate. It was shown that 2 % of fluorocontaining additive provides satisfactory physicomechanical and rheological characteristics.

The catalysts were more effective with the new binder (P/NC=1.3), compared to the previously-used one (P/NC=0.8). The following combustion catalysts were used: copper-lead phthalate both pure and combined with nickel subcarbonate, as well as $CdCO_3$, $PbCO_3$ and their mixtures of various ratios. The above-mentioned additives increase the burning rate by a factor of 1.4-1.8 within the whole interval of pressures investigated times, though, there is no significant decrease of the coefficient ν.

The potassium ferricyanide contains which is an oxidizer, was used in the composition of the propellant. Besides, it contains a cobalt catalyst (13 %).
It is necessary to add 61 % of this filler to ensure α=l. For such a propellant, the value of ν is 0.4 under a pressure in the range of 2-10 MPa.
In this way, the investigations carried out demonstrated that there could be a real possibility of the development of double based propellants, which would have high oxygen balance (α=l) at relatively low combustion temperature and provide a little (approx. l %) content of combustible gases in the chamber and their full absent at the nozzle end.

REFERENCES

1. B. P. Zhukov: Powder, pyrotechnics and special solid propellants used to fight fires. in: "Effective systems of fire-suppressing on the basis of powders and special solid propellants". Moscow, 1991, p.4-10.
2. N. P. Kopilov: New developments of fire-preventing techniques. in: "Effective systems of fire-suppressing on the basis of powders and special solid propellants". Moscow, 1991, p.10-16.
3. V. D. Serov, V. I. Klichkov, P. A. Porohovnikov, I. V. Tartinov: Use of nitrogen generators to fight fires. in: "Effective systems of fire-suppressing on the basis of powders and special solid propellants". Moscow, 1991, p.17-19.
4. Yu. V. Frolov: Gas generating and aerosolforming during combustion applying to the problem of fire-suppressing. in: "Effective systems of fire-suppressing on the basis of powders and special solid propellants". Moscow, 1991, p.28-30.
5. McHale E. T.: Experimental evaluation of Chemical Agents in Suppressing Missile Flaine Afterburning. AIAA Paper, 1975, No.75-1233, p.5-17.
6. B. P. Perepchenko, T. P. Korobenina, G. G. Shahrai, Yu. A.Militsin: New types of propellants - sources of aerosols for the systems of fire-suppressing and preventing of the explosions of gases-air mixtures. in: "Powders and special solid propellants in national economy". Moscow, CNIINTIKPK, 1990, p.18.

REFORMULATION/REUSE OF EXPLOSIVES, PROPELLANTS AND PYROTECHNICS FOR COMMERCIAL APPLICATIONS

DAN BURCH
Crane Division
Naval Surface Warfare Center
300 Highway 361
Crane, Indiana 47522, USA

1. Abstract

The U.S. Department of Defense (DoD) faces many environmental issues and concerns in the demilitarization and disposal of obsolete, excess and unserviceable munition items. Removal technology has already been demonstrated for the majority of explosive and propellant loaded items but the reclaimed energetic material is being destroyed by OB/OD or other destruction technologies. Increased concern for the protection of the environment has mandated that DoD must pursue improved methods of munitions demilitarization while maximizing ingredient recovery. For the past few years, the Naval Surface Warfare Center, Crane Division (NAVSURFWARCENDIV), Crane, Indiana has concentrated on reclaiming the valuable energetics with subsequent reformulating for commercial applications.

Efforts under the Small Business Innovation Research (SBIR) program have demonstrated the feasibility on a lab/bench scale the use of reclaimed military explosives as commercial blasting agents (metal brazing and commercial mining). Various reclaimed PBXs and other explosives have been reformulated and tested to produce an explosive of equal or superior performance as the currently used commercial explosives. Work has also been initiated to recovery RDX and HMX from military explosives and propellants and qualification of the RDX and HMX for commercial applications. Work completed under a Phase I SBIR indicates that a feed supplement for ruminant animals and a slow nitrogen release fertilizer can be generated from surplus Navy gun propellants. The feasibility of using surplus gun propellants in a novel oil and gas well stimulation process was also demonstrated.

NAVSURFWARCENDIV Crane has performed work on recovering powdered metals and other valuable constituents from pyrotechnic devices, i.e. illuminating signals, submarine location markers, photoflash cartridges and decoy flares.

H. Krause (ed.),
Conversion Concepts for Commercial Applications and Disposal Technologies of Energetic Systems, 113–120.

2. Introduction

The Navy currently has in excess of 20,000 short tons of ammunition in the Munitions Disposal Account "demil account" which is either obsolete, unserviceable or excess. The Single Manager for Conventional Ammunition (SMCA) has over 400,000 short tons and this number is projected to increase substantially in the next few years due to the world peace movements and the downsizing of the military. The primary means currently used for disposal is OB/OD which is facing close scrutiny by Federal and State environmental regulatory agencies, environmental interest groups and the public. Region IV EPA and some states have notified DoD activities that technology must be implemented to replace OB/OD or the Subpart X Part B permits will not be issued. Research is required to develop alternative technologies to OB/OD while maximizing recovery of the valuable ingredients contained in the military munitions.

The pursuit of a replacement for OB/OD has been the basis for the Navy's Ordnance Reclamation Program. NAVSURFWARCENDIV Crane in support of the Naval Sea Systems Command has pursued reclamation of propellants explosives and pyrotechnics in military munitions. In the past, several technologies have been developed for recovery of ingredients from pyrotechnic ingredients such as illumination signals, decoy flares and photoflash. NAVSURFWARCENDIV Crane through contracts with the Rock Mechanics and Explosive Research Center, University of Missouri-Rolla (UMR) have demonstrated the capability to washout PBX explosives from Navy projectiles and warheads utilizing high pressure waterjet technology. Development work is currently on-going to demonstrate the feasibility of using high pressure waterjets to remove Class 1.1 propellants from rocket motors. The Navy has broken down 5" to 16" Propelling Charges containing single, double and triple base propellant for several years with unsuccessful attempts to market the propellant. The propellant was ultimately disposed of by open burning. The efforts by the Navy have concentrated on development of technology to reformulate and reuse the reclaimed explosives and propellants.

3. High Pressure Waterjet Washout

For many Navy warheads, bombs, and projectiles containing Plastic Bonded Explosives (PBX), steam and hot water washout are not effective for the removal of the explosive from the ordnance hardware. Methods have been developed by the Navy through efforts coordinated by NAVSURFWARCENDIV Crane to demilitarize PBX loaded munitions without creating undesirable pollutants and with provisions for reclaiming reusable materials. The University of Missouri-Rolla (UMR) has investigated the use of high pressure waterjets for removal of the PBX from ordnance. Research has been directed along two significant courses. First the safety of the operation was thoroughly investigated, while the second part of the effort concentrated on the design of an automated facility.

The potential use of high pressure waterjets in an explosive environment dictated that an accurate measure of the safety be established. To determine the sensitivity of the different PBXs to high pressure waterjet impact, a test method was developed at UMR that identified two levels of reaction. The first level was that at which any detectable reaction occurred on the explosive surface and the second consisted of the level of impact at which a sustained reaction occurred. Of the Navy and Air Force PBXs evaluated, reactions were rarely detected below impact velocities on the order of 1.5 to 2.0 km/sec. Such an impact will generate pressures on the order of 160,000 to 250,000 psi. The duration and magnitude of the reactions increased, as the impact velocity was increased. At a jet impact velocity of 6 km/sec, which gives a sustained pressure of 2.25 million psi, the samples virtually all detonated. Evaluations of the removal of the different PBXs demonstrated that operating jet pressures of 20,000 psi or below were sufficient to remove all of the PBXs of interest.

As the safety evaluation progressed, the results were sufficiently promising to justify development of a vehicle for the purpose of removing the PBXs from Navy warheads. The equipment developed[1] as given the acronym WOMBAT (Waterjet Ordnance and Munitions Blastcleaner with Automated Tellurometry).

The development of the WOMBAT has progressed to provide additional flexibility in cleaning a broader variety of munitions. Procedures have been developed for removing explosives from munitions with complex internal configurations. Items washed out to date include: SPARROW, SHRIKE and HARM Warheads as well as various simulated warheads. The evaluation of washout procedures with the WOMBAT on ordnance items with differing configurations has proven that high pressure waterjets can be a viable method of PBX reclamation.

Based on the success of the PBX washout program, NAVSURFWARCENDIV in support of the Joint Service Large Rocket Motor Disposal Program at the U.S. Army Defense Ammunition Center and School, Savanna, Illinois has contracted with UMR to develop high pressure waterjet technology for washout of Class 1.1 Rocket Motors. Comparing the data generated from the waterjet impact sensitivity and the waterjet parameterization (cutting analysis) show that an acceptable margin of safety exists for washout of Navy, Air Force and Army propellants from the rocket motors. Work is ongoing in assessing all safety and environmental requirements in handling the process streams to develop operational parameters.

4. Explosive/Propellant Reuse

Since high pressure washout has proven to be a viable process for separation of PBX from the munitions hardware, a process is needed for subsequent recovery of the explosive components. NAVSURFWARCENDIV Crane is currently looking at potential commercial uses for the washed-out explosive.

The success of two separate Small Business Innovation Research (SBIR) contracts have demonstrated the feasibility of reformulating the reclaimed PBX for two different commercial applications. Work under these contracts is ongoing and shows excellent potential for a use/market for the washed out explosive. The conversion of demilitarized propellants and explosives into a commercial explosive offers a potentially cost effective process of disposal/reclamation. This concept is advantageous to the Government because it eliminates the environmental disposal problem by utilizing all of the components of the waste stream.

4.1. COMMERCIAL MINING EXPLOSIVE

There is in excess of 5 billion pounds of explosive used annually in the mining industry. Presently, the use of ammonium nitrate and fuel oil (ANFO) as an explosive is advantageous to many industries because of price and the relative insensitivity of the material. Utilization of more effective explosives which could be made from the same constituents found in propellants and PBXs would not normally be considered due to the high costs of the material. However, since the material in question is a waste product that would be disposed of by destruction, the possible use becomes much more conceivable to potential users. Consequently, not only has a use been found for the disposal assets, but the Government has eliminated the environmental problems associated with disposal.

Technology Development Inc. (TDI), Rolla, Missouri is addressing the market potential for the reclaimed explosives and propellants in the commercial mining industry. The use of reclaimed PBX explosive offers a tremendous potential to the commercial mining industry because the converted explosive could reduce drilling costs associated with the blasting operations. This is accomplished by developing a more effective explosive which could reduce the diameter and/or the relative spacing of the boreholes.

TDI demonstrated in the Phase II SBIR contract[2] that a commercial blasting agent can be reformulated from the washed out explosives with equal or better performance than ANFO. Microformulation studies showed the PBXs to be deficient in oxygen. TDI developed a pilot scale manufacturing plant capable of producing 100 pounds of reformulated explosive per day and installed the unit in an underground mine at UMR. The pilot plant consisted of a vibratory screen, a vacuum dryer and a stick packing machine. Two to four pound explosive sticks were produced from several PBX/oxidizer formulations and subjected to velocity of detonation determinations in the UMR test quarry. Representative detonation velocities for selected explosives and comparison with commercial explosives are reported in Table 1.

Table 1

Explosive	Density (g/cc)	Velocity of Detonation (km/sec)	Condition
ANFO	0.81	3.66	Confined
Emulsion high explosive	1.19	5.79	Confined
Ammonia gelatine dynamite	1.40	6.10	Confined
Composition B	1.60	7.93	Confined
PBXN-3/AN	0.90	3.39	Unconfined
PBXN-4/Sodium perchlorate	1.06	5.18	Unconfined
PBXN-101/AN	0.99	5.79	Unconfined
PBX-105/AN	1.33	5.71	Unconfined
PBXN-101/AN/Urea	0.9	4.33	Unconfined

The limited velocity of detonation results obtained did confirm that an effective mining explosive can be produced from the reclaimed military explosive. Plans are to perform additional tests on different explosive and Class 1.1 propellant formulations and produce sufficient explosive for qualification at local mines and quarries.

4.2. COMMERCIAL METAL BRAZING EXPLOSIVE

TPL is exploring the use of the reclaimed explosive for use in metal bonding applications. Explosive Fabricators Inc. (EFI), Louisville, Colorado alone uses in excess of two million pounds of explosives a year in this process to clad dissimilar metals. EFI requirements are that the explosive have a detonation velocity of 2.2 km/sec with a velocity variability of ±50 m/s.

TPL conducted a series of product acceptance tests to confirm that the vendor's defined detonation velocities and repeatability/variability are achievable with the reclaimed PBX/diluent mixes. Tests were also ran with the selected explosive/diluent mix with the explosive currently being used, an ANFO/perlite mix. TPL, employing an improved mechanical reprocessing method for the PBX ran a test with mix of 36 wt% PBX(AF)-108/64 wt% diluent. The results[3] employing a detonation velocity tray test configuration of 12"w x 18"l x 5"d follow:

Table 2

Wire Pair	Detonation Velocity (km/s) Shot 1	Shot 2	Shot 3
1-2	2.19	2.21	2.22
2-3	1.81	2.24	2.22
3-4	2.21	2.26	2.02
4-5	2.36	2.33	2.22
5-6	2.64	2.24	1.85
1-6	2.24	2.25	2.12

The average detonation velocity obtained for the three tests was 2.20, and excellent test-to-test and within-test uniformity were obtained in the TPL detonation velocity testing at the 5" depth with the 36/64 PBX to diluent mix.

A 50/50 mix was prepared and transported to EFI where the desired 36/64 mix was prepared. An 18"w x 36"l x 5"d detonation velocity tray test was conducted by EFI to verify conformance with their requirements. The results obtained by the TPL measurement technique at EFI were comparable to the results obtained at TPL and the velocity variation was less that the ANFO/perlite mix.

Two explosive bonding tests were then performed by EFI to evaluate the PBX(AF)-108 blasting agent. The bonding test consisted of 3/8" stainless steel bonded to 1 3/8" carbon steel. EFI employs this combination as a reference material. The carbons steel plate and the stainless steel plate had dimensions of 22"w x 36"l. The results obtained from the initial bonding experiments showed outstanding interface and bonding characteristics between the carbon and stainless steels. Also an added advantage is the detonation velocity of the TPL explosive can be adjusted to that required for the explosive bonding applications. TPL plans to build a pilot plant manufacturing unit in order to produce sufficient reformulated explosive to provide the metal fabricator, where full scale detonation velocity measurement tests will be performed on the tailored, reclaimed PBX material.

Also during the Phase II SBIR effort, TPL developed a multiple use demolition explosive from the washed-out PBXN-108 which showed promised for use as a non-combat explosive. Excavation experiments were performed comparing the performance of the granular PBXN-106 with that of ANFO. The resulting craters were conical in shape and 32.5 pounds of the PBXN-106 resulted in 104 ft^3 of earth removal compared with 78.5 ft^3 removed of earth removed by 45.0 pounds ANFO. Tests were also conducted by the Demolition Program Office at the U.S Army Armament Research, Development, and Engineering Center which demonstrated the feasibility of using the granulated PBX as a demolitions explosive.

4.3. SURPLUS GUN PROPELLANT FOR COMMERCIAL APPLICATIONS

One of the largest generations into the disposal account is gun propellant (single, double and triple base propellants). TPL under a Phase I SBIR contract has conducted preliminary experimentation and shown feasibility of using extracted ingredients from the nitrocellulose based propellants for agriculture applications[4]. Tests performed at New Mexico State University have demonstrated the concept of using the nitrocellulose recovered from single base propellant as an animal feed supplement and the nitrocellulose/nitroguanidine recovered from triple base propellant as a specialty fertilizer.

In vitro rumen culture tests demonstrated the ability of unadapted rumen microbes to digest nitrocellulose. The digestibility of the nitrocellulose was similar to that of cottonseed meal, an accepted animal feed supplement. Feeding of a sheep with nitrocellulose showed an increased food consumption. Based on the nitrocellulose digestion rate measured, the use of the nitrocellulose as an animal feed supplement could result in a consumption rate of over one million pounds per year just from New Mexico feedlot usage alone. Nitrogen release to sorghum was demonstrated by greenhouse pot studies for single base propellant, nitrocellulose, extracted nitroguanidine and a triple base propellant water extract. A Phase II SBIR contract was recently awarded to TPL, Inc. to continue the development effort.

TPL also demonstrated by using a tailored pulse fracturing technique the use of gun propellant as a specialty gas generation device for oil and gas well stimulation. This work is continuing under a DOE Phase II SBIR contract and will include testing of Navy propellant.

4.4. RDX/HMX RECOVERY

NAVSURFWARCENDIV White Oak is currently exploring new extraction methods and solvents to recover RDX and HMX from PBX explosives and Class 1.1 propellants. Specifically, alternative solvents to dimethyl sulfoxide are being examined as well as extraction with more volatile solvents under increased temperatures and slightly increased pressures (1-5 atmospheres). Besides exploring the use of the non/low hazardous organic solvents in the pure state, azeotropic mixtures will also be looked at for solvolysis of the RDX and HMX. The solvent butyrolactone shows promise for extraction of HMX from reclaimed explosive materials. Trade-offs between efficiency, environmental and industrial hygiene concerns are being explored which may make several solvents attractive alternatives to DMSO.

TPL, Inc. in a Phase I SBIR contract has been conducting experiments to recover HMX from Class PBX explosives and propellants. A process developed using a dilute hydrochloric acid was effective in recovering HMX from PBXN-3 (nylon binder). TPL has recovered sufficient quantity of the HMX to provide to commercial users (perforating charges for oil and gas wells), who have shown interest in the reclaimed HMX.

5. Pyrotechnic Recovery

NAVSURFWARCENDIV Crane has performed research in the past in recovering powdered metals and other valuable constituents from pyrotechnic devices, i.e. illuminating signals, submarine location markers, photoflash cartridges and decoy flares. Some of the key solvents used in the separation techniques are ozone depleting chemicals and can no longer be used. Research has been initiated to find substitute solvents or alternate technologies to recover the valuable ingredients. Also laboratory investigations have been initiated to develop processes for recovery ingredients from some of the new pyrotechnics smokes and flares. Investigations will continue on the bench/pilot scale to develop removal, recovery or chemical conversion technologies to reclaim usable and valuable ingredients contained in the pyrotechnic formulations.

6. Conclusion

For many munition items in the Navy's "demil account," there is no available disposal capability other than OB/OD. Research efforts have demonstrated on the laboratory/bench and pilot scale the ability to recover valuable ingredients or reuse the constituents from many Navy explosive and propellant items. Efforts to date indicate that an effective explosive can be formulated from the reclaimed material which could provide benefits to potential users plus providing a market for the reclaimed energetic material.

The development of these reclamation technologies will allow DoD to eliminate the dependence on OB/OD and continue to reduce the demil backlog. The technologies will allow for recovery and reuse of the valuable ingredients thereby reducing hazardous waste generations. The long term result will be to free valuable magazine storage space at the DoD activities which is becoming critical at many Navy and SMCA locations.

7. References

1. Craig, L., Short, J., Summers, D., Worsey, P. (1986) Evaluation of Washout Procedures for Missile Casings, Report No. NWSC/CR/RDTR-291.

2. Wulfman, D. and Nixon, F. (1993) Environmentally Neutral Reformulation of PBX Explosives, Report for NSWC Contract N00164-91-C-0055.

3. Stoller, H., Perry, M. (1992) The Development of a Reclaimed PBX Product for Commercial Applications, Report No. TPL-MR-1000-21/22 for NSWC Contract N00164-91-C-0028.

4. Stoller, H., Kidd, et al, (1992) A Study of the Reuse/ Reformulation of Navy Gun Propellants", Report No. TPL-FR-2023 for NSWC Crane Contract N00164-92-C-0021.

PYROTECHNIC AEROSOL GENERATING COMPOSITIONS AND FIRE SUPPRESSING SYSTEMS ON THEIR BASIS

N. A. Silin, V. N. Emel'janov

Scientific Research Institute of Applied Chemistry,
141313 Sergiev Posad, Moscow Region, Russia

Until recently different types of Halons have been most effective and widely used fire extinguishants all over the world. However, these agents affect the Earth ozone layer adversely and in line with international agreements (Montreal, 1987; London, 1990) their production and use are decreased and will be completely stopped soon.

Investigations carried out at the Institute of Applied Chemistry in 1963-64 and recently have demonstrated that Halon systems could be well replaced by devices with pyrotechnic aerosol generating compositions.

Present-day fire extinguishants and systems having been analysed, some main requirements for pyrotechnic aerosol generating compositions were developed.
They are:

- high fire suppressing power;
- non-toxic aerosol generation products;
- environmental safety;
- minimum burning temperature;
- chemical and storage stability;
- large domestic sources of raw materials;
- combustion products must not cause the corrosion of structural materials.

H. Krause (ed.),
Conversion Concepts for Commercial Applications and Disposal Technologies of Energetic Systems, 121–129.

Due to the high particle dispersity aerosol fire suppressing power and sedimentation stability last 25-30 min:

AGC 51-35-1 Specification	
burning rate, mm/sec	2,5 - 3,0
combustion temperature, °C	800-1000
slag fraction, %	0,6 - 1,0
sensitivity:	
impact	low
class	19
friction	medium
class	13
electric spark	less than 20 mJ.

The development and investigation of aerosol generating compositions were accompanied by constructing fire suppressing aerosol generators.

The weight of generator AGC charges available ranges from 0,2 to 2,6 kg.

Main fire suppressing generator characteristics are given in Table 4, test result - in Table 5.

As can be seen from Table 5, fire suppressing aerosol generators (FSAG) have confirmed their effectiveness with several full-scale objects and mock-ups differing from one another by configurations, volumes, equipment amount and type.

Besides, fire suppressing aerosol systems are characterized by smaller sizes and lighter weights as compared to Halon devices.

Thus the FSAG weight (for object 2 in Table 5) was as little as 32,2 kg while that of Halon 1301 system amounted to 263 kg and of Halon 2402 system - to 305 kg, i. e. the weight was almost 10 times larger.

FSAG generators are made from materials and units widely used in pyrotechnics.

At present different industrial locations and residential premises are equipped with FSAG generators described above.

Studies conducted at the Institute have shown that pyrotechnic aerosol generating compositions based on alkali metal nitrates and organic nitrogenated compounds meet the above requirements most completely.

By now some formulas of aerosol generating compositions (AGC) developed at the Institute of Applied Chemistry are available:

- 51 - 34 type copmpositions for space fire extinguishment inside uninhabited closed or semiclosed (vented) chambers;
- non-toxic 51-35 type compositions for space fire extinguishment inside inhabited closed premises.

The above aerosol generating compositions have undergone extensive laboratory / bench tests and received several certificates from some organizations confirming:

- the lack of ozone destructive action (the Institute of Energetic Problems in Chemical Physics, the Russian Academy of Scienes);
- the manner of combustion products influence on a electronic equipment (the "Energy" Scientific Industrial Association);
- corrosion resistance of structuraal materials (the Institute of Aviation Materials);
- non-toxicity of the described composiotions (the Toxicology Institute, the Ministry of Health).

Aerosol generating compositions proved to be more effective than others.

Aerosol generating compositions and fire suppressing aerosol generators developed at the Institute are protected by 12 USSR author certificates and by one Russia patent.

It should be noted that FSAG generators can be actuated both by an electrical signal from the system control panel and by an independent thermal sensor. The latter is based on a pyrotechnic current generator and comes into action when the temperature of a protected object rises above the established threshold level.

The independent thermal sensor is characterized by

– outside diameter, mm	60
– height, mm	110
– temperature operation threshold, °C	72 ± 5
– response time, sec	no more than 90
– maximum current strength in a pulse under the load of 4,0 ohms, A	no less than 2,0

Test results are given in Tables 1,2,3.

Table 1

Comparative results of testing AGC 51-35-1 and Halon 2402 during fire extinguishing different materials inside the

volume 0,25 M^3 chamber

Combustible material	Fire extinguishing concentration, g/m^3	
	AGC 51-35-1	Halon 2402
Acetone	30-50	180
Heptane	35-50	190
Ethanol	40-50	210
Polyurethane foam	75-90	300
Wood	100-140	320

Table 2

Maximum (peak) desensitizing concentrations of fire extinguishants for combustible gas - air mixtures

Fire extinguishant	Peak desensitizing concentration, g/m^3	
	Methane-air mixture	Hydrogen-air mixture
Nitrogen	520	80
Halon 2402	325	1700
Halon 1301	320	1600
AGC 51-35-1	50	230

Table 3

Toxicity testing results of some fire extinguishants under fire suppressing concentration (expose time 15 min)

Fire extinguishant	Toxicological effect, %	
	death, %	loss of the ability to move, %
Halon 1301	0	0
Halon 1211	0	0
Halon 2402	0	14,6
Carbon dioxide	10,4	95,8
AGC 51-35-1	0	0
STK-B	100	100
STK-5-2	0	14,3
STK-5-2,5D	79	100

Thus comparative tests have confirmed the high effectiveness of AGC 51-35-1.

When burning, the above composition produces a mixture of inert gases and ultrafine particles with a large specific surface inhibiting organic materials combustion in the atmospheric oxygen.

The aerosol disperse phase generated by the burning composition 51-35-1 having been chemically analysed, main substances proved to be:

K_2CO_3 1,5 H_2O	52,7 %
NH_4HCO_3	25,7 %
$KHCO_3$	8,2 %
KNO_2	7,9 %
other potassium compounds	5,5 %.

In this case the mass fraction of different size particles is as follows:

> 1 μm	26,6 %
1-2 μm	64,9 %
2-5 μm	8,5 %.

Table 4

FSAG GENERATOR SPECIFICATION

CHARACTERISTIC	CHARACTERISTIC VALUE					
	FSAG-26	FSAG-1.3	FSAG-1.0	FSAG-0.6	FSAG-04	FSAG-02
1. Protected volume, m^3	30-50	15-25	10-20	6-12	4-8	2-4
2. Generator mass, kg	4,6	2,7	2,1	1,4	1,0	0,64
3. AGC charge mass, kg	2,6	1,3	1,0	0,6	0,4	0,2
4. Overall dimensions, mm						
- length	308	188	215	155	120	90
- diameter	114	114	80,2	80,2	80,2	80,2
5. Generator operating time, sec	14-30	14-30	8-10	8-10	6-8	6-8
6. The generator is actuated when supplying						
- voltage, V	19-30	19-30	14-30	14-30	14-30	14-30
- current, A	1,5-3,0	1,5-3,0	1,5±0,1	1,5±0,1	1,5±0,1	1,5±0,1
- pulse duration, sec	0,01	0,01	0,01	0,01	0,01	0,01
7. Safe current (A)	0,25 + 0,01	0,25 + 0,01	0,05 + 0,0025	0,05 + 0,0025	0,05 + 0,0025	0,05 + 0,0025
during (sec)	5,0 ± 0,3	5,0 ± 0,3	300	300	300	300

Table 5

Object characteristic	Volume, m^3		Space fraction occupied by equipment, %	Combustible material	AGC consumption, g/m^3	Extinguishment time, sec
	object as a whole	equipment				
1. Electric cabinet SHM-1	1,3	0,33	25,4	Ethanol, polymethyl methacrylate	38-58	5-14
2. Electrical equipment location	260	88	33,8	Ethanol, diesel fuel	42	24
3. Do	260	5	1,9	Do	38	18
4. Part of cable tunnel	30	2	6,7	Diesel fuel, cable	90	25
5. Electronic equipment and computer location	58	17	29,3	Ethanol	35-57	23-45
6. Do	131	1,3	1,0	Do	39-51	13-20
7. Do	536	220	41,0	Ethanol, cable	88-94	18-23
8. Diesel generator location	14	2,5	10,9	Ethanol, diesel fuel	54	32
9. Diesel locomotive TEZ	113	23	20,4	Kerosene, diesel fuel	54	32
10. Ship cabin mock up	9,1	0,5	4,9	Heptane, diesel fuel	31-62	3-48
11. Ship machine room mock up	24	0	0	Diesel fuel	85,7	24
12. Do	168	1,9	1,1	Diesel fuel	85,7	24
13, Do	1300	115,1	8,9	Diesel fuel	75,8	20

CONCLUSIONS

Studies carried out at the Institute of Applied Chemistry explore the possibility to replace Halons and Halon fire suppressing systems by pyrotechnic aerosol generating compositions and FSAG generators.

Pyrotechnic aerosol generating compositions developed and FSAG generators designed make it possible to provide highly effective and envoronmentally safe fire suppressing systems for use in industry and in the home.

Due to the conversion it's possible to organize a lot production of FSAG generators at low cost and in the shortest possible time and to meet the country demands for new effective fire suppressing systems over the next 3-4 years.

USAGE OF EXPLOSION TECHNOLOGY FOR RECOVERY OF ARMOURED ORDNANCE

B. A. Abramov, B. P. Zhukov, V. G. Volohov, E. I. Shemiakin, I. B. Kremnev, I. M. Balojan, V. P. Pahomov, I. A. Potapov, D. F. Marov

Conventional Armed Forces in Europe of more than 20000 units of the armoured systems should be recovered in 1991 - 1995. A great amount of the high-quality steel will be obtained during this recovery. So the problem of economical and ecologically harmless processing of the tank cases and turrets is very vital.

Components of the case (sides, front armoured plates, bottoms) can be used as flat plates for production of billets and plate rolling etc. There are two ways of recovery of these components:

a) disintegration of the case and the turret into the large elements (3-5 parts) and
b) disintegration of the case into the small parts for metallurgic processing of the armoured systems.

The disintegration of the cases and the turrets by explosion can be produced with the help of standard recovered industrial ammunition and explosives and special shaped charges (Table 1).

The main idea of the explosion disintegration of the cases into the large parts is a fragmentation of welded seams by a shattering action of the explosion. The scheme of the explosive placement for the shattering action of the charges in the zone of the welded seams, the given size of the charges and the technique of their initiation provide the disintegration of the tank cases into the components (sides, bottom, top and bottom front plates etc.) with minimum scattering and deformation. Disintegration of the turrets with the help of the superimposed charges is efficient only for separation of the turret door and its disintegration in the longitudinal plane.

H. Krause (ed.),
Conversion Concepts for Commercial Applications and Disposal Technologies of Energetic Systems, 131–136.

Table 1

Characteristics of the ammunition recommended for the disintegration of the armoured systems by explosion			
Ammunition type	Explosive mass, kg (kg/m)*	Dimension, mm	Thickness of the cut of the steel sheet, mm
Explosive			
Trotyl	0.2	25x50x100	-
Trotyl	0.4	50x50x100	-
PVV-4	1.0	70x70x140	-
Extended charges			
S3-19	1.0*	7x50x2000	10
S3-4P	4.0*	0.45x2000	15
S3-6M (PVV-4)	6.0*	0.82x1200	40
S3-6M (PVV-5)	-	-	-
DKRP-4	-	0.70x10300	-
Shaped charges			
K3-5	8.5	-	450
K3-6	1.8	-	215
LK3-80	8.0	-	80
K3U	12.0	190x225x500	120
SHK3-4	0.6	-	6
SKH3-6	1.1	-	16

Disintegration of the cast part by superimposed charges results in the appreciable consumption of the explosives, so it is expedient to disintegrate the side-frames of the turrets by the integrated technique: by prelimenary cutting of the turret with special shaped charges and than the final disintegration with the superimposed ones. Thus the mass of the explosive charges is decreased by a factor of 2 as compared with the mass of the superimposed charges made of trotyl. As the thickness of the side cross section is about 160 mm two successive blastings of the charges of the 3KLB-120 type should be done to receive the given depth of the cutting. One blasting will be enough in the rear part of the turret with the thickness of 65 mm. The extended explosive charge is placed and then exploded inside the turret across the formed cuttings. As a result the turret is broken into four parts along the cuttings and slight cross section of the beam mask. Operation charts of the placement and consumption of the charges during disintegration of the tank cases and turrets are given in the Table 2 and 3.

The total mass of the explosive needed for the case disintegration into the large parts is from 50 to 60 kg for different tanks and that for the turrets is about 50 kg. Labour consumption of the operations for the disintegration of the tank cases and turrets is about 4.0 and 2.0 man-hours. In order to disintegrate the tank case into small parts (1200x1000 mm) the 3 KLB-90 charges are exploded in the areas of the expected cuttings of the side, nose and stern plates that results in the formation of the cuttings and in weakening the given cross sections. Then the technology of the tank disintegration is repeated completly in addition the action of the shock wave from the inside results in demolition of the plates along the cuttings with simultaneous breaking of the welded seams.

The shaped charges of the SH3 type can be used for disintegration of the infrantry combat vehicles and other armoured combat vehicles with the thickness of the walls up to 20 mm. The SH3 charges are placed along the line of the anticipated cuttings on the outside wall of the case. Thus the disintegration is received by one blasting without scattering of the fragments. Labour consumption of the operations for the disintegration of the infantry combat vehicle is 8 man-hours, consumption of the SH3 charges is 37.5 kg.

Table 2

Consumption of the 3KLB-60 and 3KLB-90 charges for disintegration of the tank into the structural elements.				
Operations and elements	**Cutting amount**	**Cutting length, m**	**Charge mass, kg 3KLB-60**	**3KLB-90**
Case disintegration into elements.				
1. Bottom -sides	2	8.86	11.5	-
2. Bottom -stern	1	1.60	2.1	-
3. Bottom -side -boxes	2	1.70	2.3	-
4. Side -turret boss	2	2.00	2.7	-
5. Side -stern	2	0.90	1.2	-
6. Side -nose plate	2	1.10	1.3	-
7. Turret plate nose plate	1	1.94	3.5	-
8. Top nose plate lower nose plate	1	1.82		9.6
Additional cuttings				
1. Top nose plate	1	1.22	-	6.4
2. Lower nose plate	1	0.65	-	3.4
3. Sides	4 2 6	1.40 0.90 4.50	- - -	7.4 4.7 23.7
4. Stern plate	1	0.75	-	3.9
Disintegration of the case elements into parts				
1. Bottom	4 1 1	7.27 2.10 5.10	14.5 4.5 10.5	- - -
2. Lower side plate		0.65	1.4	-
3. Turret plate	1	0.75	1.6	-
Disintegration of the turret				
1. Door	1	6.65	17.5	-
2 Upper cross section	1	0.45	-	2.3
3. Stern cross section	1	0.90	-	4.7
4. Mask beam	1	0.15	-	0.7
5. Side cross section	4	3.20	-	16.8
		Total charge mass	74.9	83.6

Table 3

Operation chart of the consumption of the 3KLB-60 and 3KLB-120 charges during disintegrationof the tank turret						
Area of the charge installation	**Cutting amount**	**Cutting length, m**		**Charge mass, kg**		
		One	**Total**	**3KLB-60**	**3KLB-120**	**Trotyl**
1. Turret cover						
- profile of the welding to the turret	1	5.30	5.30	13.9	-	-
- center seam	1	1.35	1.35	3.6	-	-
- top cross section of the turret	1	0.45	0.45	-	2.3	-
- stern cross section	1	0.90	0.90	-	4.7	-
- front beam of the mask	1	0.15	0.15	-	0.7	-
- side cross section	2	0.80	1.60	-	8.4	-
- side cross section	2	0.80	1.60	-	8.4	-
- side cross section	2	0.80	1.60	-	-	8.0
- side cross section	1	0.90	0.90	-	-	4.5
			Total 54.5 kg			

Gas cutting is an alternative technique of the case and turret disintegration which is widely used in the industry.

Gas cutting has the following advantages:

- wide assimilation in the industry;
- production of the equipment and materials;
- availability of highly qualified specialists;
- feasibility of cutting to the depth of 400 mm.

Its disadvantages are:

- high cost of the materials;
- impossibility of cutting the structures containing organic fillers and aluminium and titan structures;
- contamination of the environment;
- high fire hazard when cutting the structures contaminated by fuel or oil.

In comparison with the gas cutting the explosion disintegration has following advantages:

- low labour consumption;
- feasibility to use recovered explosives and ammunition;
- feasibility to remove organic and nonorganic fillers out of the armoured systems;
- less ecological hazard;
- feasibility to use this technique in the removed regions.

Disadvantages of the explosion technique:
site with special equipment and specialists is required.

Comparison of the economic aspects of two above techniques is given in Table 4.

Table 4

Comparative economic characteristics of the operations of the tank cases and turrets disintegration by gas and explosion cutting									
Operations	**Gas cutting characteristics**					**Labour consumption, man-hour**		**Cost**	
	Plate thickness, mm	**Consumption, 1/min oxy-gen**	**ace-tilen**	**Cutting rate, mm/min**	**Cutting length, m**	**Expl. tech-nique**	**Gas cut-ting**	**Expl. tech-nique**	**Gas cut-ting**
1. Disintegration of the case into elements						4		1.0	1.73
-turret plate	15	224	34	435	3.0	-	0.12	-	-
-bottom	20	292	45	395	15.0	-	0.64	-	-
-stern	45	640	84	235	4.8	-	0.34	-	-
-side	80	1200	130	130	18.4	-	2.36	-	-
-front plate	100	1500	158	158	18.4	-	1.64	-	-
2. Turret disintegration									
	160	2970	220	94	5.3	2.0	0.94	1	0.83
3. Demolition						7.06		1.0	2.04
-turret plate	15	224	34	435	4.0	-	0.15	-	-
-bottom	20	292	45	395	22.0	-	1.93	-	-
-stern	45	640	84	235	5.5	-	0.39	-	-
side	80	1200	130	130	28.6	-	3.67	-	-
-front plate	100	1500	158	107	12.3	-	1.92	-	-

As we see in Table 4 the cost and labour-consumption of the explosion technique is lower than that of the gas cutting. Moreover the explosion technique permits to remove the organic fillers from the composition of the case and turret which are ecologically hazardous during the gas cutting and also nonorganic fillers which can decrease the quality of the smelting.The explosion technique can be used succesfully during the recovery of the other parts of the armoured systems, for example, trucks, caterpillars etc., which are made of different materials (metal-rubber). So the rubber bumper of the truck of the T.-64 tank can be cutted from the aluminium push disk and the truck can be broken into elements by blasting of the SH3 N2, SH3 N4 charges (their masses are 130 g/m and 340 g/m). A huge tyre of the T-55 tank truck is cutted by the blasting of the SH3 N4 charge. Its labour-consumption as compared with the other techniques (mechanical dismounting, disintegrationby thermal heating etc.) is significantly lower. Special devices are not required. However it is seen from the Table 4 that the gas and plasma cuttings are the cheapest techniques for disintegration of the case and turret elements into small parts. In this case it is possible to use stationar, automatic and semiautomatic highly efficient gas cutting equipment of the metallurgical industry.

CONCLUSION

Processing of the recovered tank cases and turrets into large metal scrap by the explosion technique is economically efficient and ecologically safe as compared with the other techniques.

The explosion disintegration is economically more efficient due to usage of the obsolete and recovered explosives.

The explosion technologies can be used successfully in recovery of the tank elements containing different materials and fillers.

PROBLEMS OF TROTYL (TNT) PROCESSING INTO CONDENSATION MONOMERS, POLYMERS AND DYES

V.A. Tartakiovsky, S.A. Shevelev, M.D. Dutov, and A. Kh. Shakhnes[1]

A.L. Rusanov, L.G. Komarova[2]

A.M. Andrievsky[3]

Presently in a number of countries, including Russia, a large amount of ammunition liable to liquidation has been accumulated. Its major explosive component is 2,4,6-trinitrotoluene (trotyl. TNT), and thus its utilization has become the basic challenge.

One of the most prospective directions in the utilization of extracted TNT, in our opinion, is a possibility of its application as a cheap multitonnage chemical source in manufacturing products with valuable technical and commercial properties. The same conclusion was made by German [1] and American [2] researchers.

[1]N.D. Zelinsky Institute of Organic Chemistry, 117913 Moscow, Russia

[2]A.N. Nesmeyanov Institute of Organo-Element Compounds, RAS 117913 Moscow, Russia

[3]MNPO, Scientific Research Institute of Semiproducts and Dyes, 103787 Moscow, Russia

H. Krause (ed.),
Conversion Concepts for Commercial Applications and Disposal Technologies of Energetic Systems, 137–149.

The present paper reviews some results of the complex investigation of the chemical conversion of TNT performed within in collaboration of N.D. Zelinsky Institute of Organic Chemistry RAS (initial TNT processing, synthesis of monomers and semiproducts) with A.N. Nesmeyanov Insitute of Organo-Element Compounds RAS (synthesis of polymers and study of their properties) and MNPO „Nauchno-Issledovatelskii Institut Poluproduktov i Krasitelei" (synthesis of dyes and study of their properties).

The objective of the investigation is the production of new TNT-based condensation monomers and polymers, as well as new dye stuffs possessing a complex of valuable technical and commercial properties.

1. Synthesis of condensation monomers and semiproducts for dyes

To obtain monomers and semiproducts for dye stuffs of the given structure on the basis of TNT and its initial processing products directed synthesis of aromatic benzoid di- and polyamines, including N-substituted, nitramines, derivatives of nitro- and aminophenols, multiatom phenols, aromatic ethers and their thioanalogs has been elaborated. Synthesis of such structures of the basis of TNT and its derivatives is attainable through directed substitution of nitro groups for necessary fragments, selective reduction to respective amines, selective transformation of reduction products, and conversions of the methyl group.

i) TNT reduction and reduction product conversions

An efficient electrochemical method of TNT reduction to 2,4,6-triaminotoluene (TAT) without any organic solvents was developed by Profs. V.P. Gultyai and V.N. Leibzon in cooperation with the authors.

Chemical and electrochemical techniques of selective TNT reduction allowing to secure not merely mono- and diamines 1 and 2, available before, but also dramatically less accessible mono- and diamines 3 and 4 with an ortho-position of amino groups vs. the methyl group were developed. 2,6-Diamino-4-nitrotoluene (4) was for the first time produced directly from TNT and through one step only.

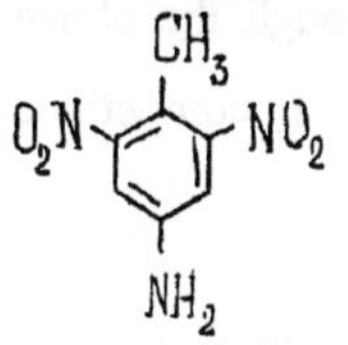
1

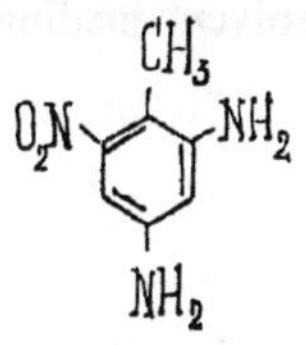
2

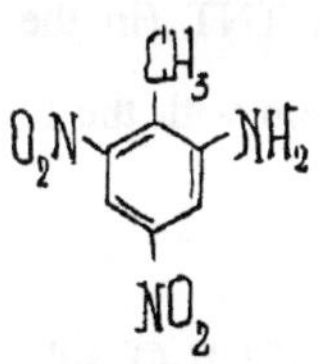
3

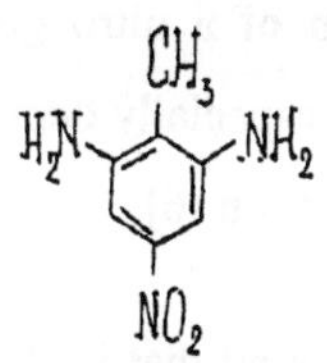
4

Diamine 4 is of interest since on its basis diisocyanate can be obtained which, by analogy with diisoyanate from diamine 2 [2], is applicable to the synthesis of polyurethanes where curing catalysts are not required.

A convenient procedure of TNT hydrolysis with the formation of 2,4,6-trihydrotoluene (methyl floroglucin), a probable floroglucin alternative, was worked out. The latter is widely applied in industry, the manufacture of dyes included [3].

Selective acylation of TAT by one amino group was performed for obtaining condensation monomers - TAT derivatives with two free amino groups.

ii) Substitution of nitro groups in TNT and products of its initial processing. Synthesis of aromatic diamines with TR substituents.

One of the main directions of the present works is creating condensation monomers - aromatic bezoid diamines containing RO-type substituents, the ones of the aromatic kind (ArO), being most preferable for improving properties of polymers. The introduction of such substituents can be achieved through nucleophilic substitution in nitro groups in TNT or its derivatives under the action of corresponding alcoholates or phenolates. Yet, it is known [4,5] that in the interaction of TNT with bases, instead of substitution of nitro groups, the formation of stable anion σ-complexes due to nucleophile addition to the aromatic nucleus is observed and a proton splitting off from the methyl group yielding 2,4,6-trinitrobenzyl anion (TNBA) with subsequent transformations of these products takes place.

At the same time theoretical (quantum chemical) calculations (PM3 method) done in collaboration with Dr. V.N. Solkan have shown that there is still a possibility of nucleofilic substitution of a nitro group in TNT (in the polar solvent medium) though the above processes, especially deprotonation with the formation of TNBA, are strongly competitive with subsitution [6].

We have found that in the case of CH_3O^- substitution of the nitro group in TNT does not occur at all. As for less basic PhO^-, the main reaction routes are characterized by some other processes although a small amount of a nitro group substitution product is yielded. In the case of thioanalogs, RS^-, the reaction proceeds more smoothly.

Since the main process that impedes substitution of the nitro group in TNT for the RO fragment is deprotonation of the methyl group, in order to avoid such an undesirable reaction course we turned from TNT to the product of its demethylation - 1,3,5-trinitrobenzene (TNB).

TNT demethylation was well elaborated previously and a series of its varients is technological acceptable

It was known before our works that such O-nucleofiles, as CH_3O^- and $PhCH_2O^-$ were able to substitute a nitro group in TNB [7,8,9].

We have worked out a method for obtaining 3,5-dinitroanisole [8] and 3,5-dinitrophenol from it.

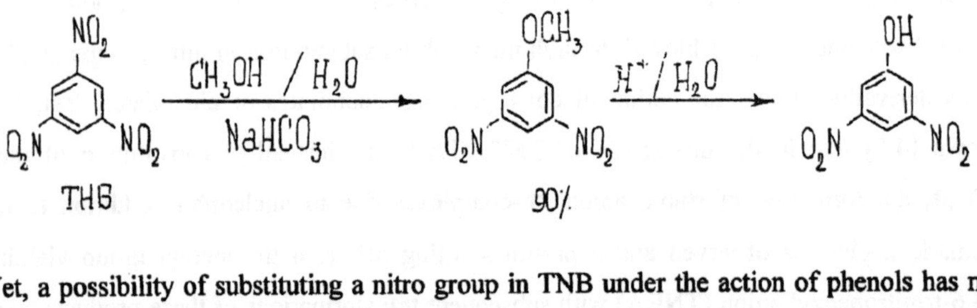

Yet, a possibility of substituting a nitro group in TNB under the action of phenols has not been virtually studied so far. It was merely known that phenolate-ion with TNB yielded anionic O- and C-σ-complexes [4,5].

We found specific conditions under which during the interaction of phenols and their analogs with TNB in the presence of bases substitution of the nitro group for the aroxy group took place, corresponding aryl ethers being formed in preparative yields. This allowed us to develop a general method for producing 3,5-dinitrophenylaryl ethers [10] that had been unaccessible earlier. In so doing bis-3,5-dinitrophenyl ethers an be also produced (originating from diatomic phenols or bis-phenols).

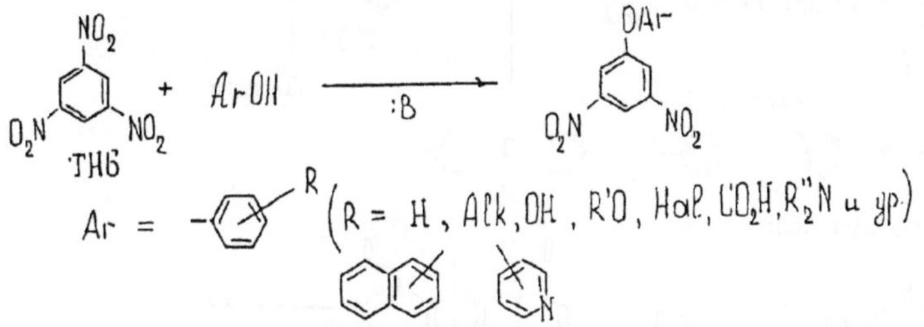

It has been found that substitution of two nitro groups in TNB for aroxy groups with the formation of earlier unknown 5-nitroresorcin aryl esters in also feasible.

Methods for reducing the obtained 3,5-dinitrophenyl ethers both to respective diamines and amino nitro derivatives were developed.

We elaborated optimum procedures including catalytic hydrogenation (in cooperation with Profs. V.Z. Sharf and Ye.F. Litvin) for the synthesis of methoxy-and phenoxy-3,5-diaminobenzenes (MDAB and PDAB, respectively) as the most accessible monomers. They allow to form these diamines with a practically quantitative yield and a high degree of purity.

2. Condensation polymers on the basis of TNT derivatives

Before we started our work the problem of producing TNT-based polymers had been restricted to the idea of utilizing TAT, the product of TNT reduction, for the synthesis of 2,4,6-trisisocyanatotoluene and 2,4,6-trismaleimidotoluene as possible monomers. Besides, TAT had been suggested to be used for producing aminoformaldehyde resins [1].

Within our studies an alternative approach to the use of TAT in the production of polymers, namely polyamides and polyimides containing benzamide and N-phthalimide arrangements, respectively, was discussed. To produce these polymers TAT and its trihydrochoride (TAT 3HC1) were introduced in the reaction, first, with one benzoilchloride or phthalic anhydride molecule for blocking one TAT amino group, and the other two amino groups in the products of TAT monoacylation were used for polyamidation reactions.

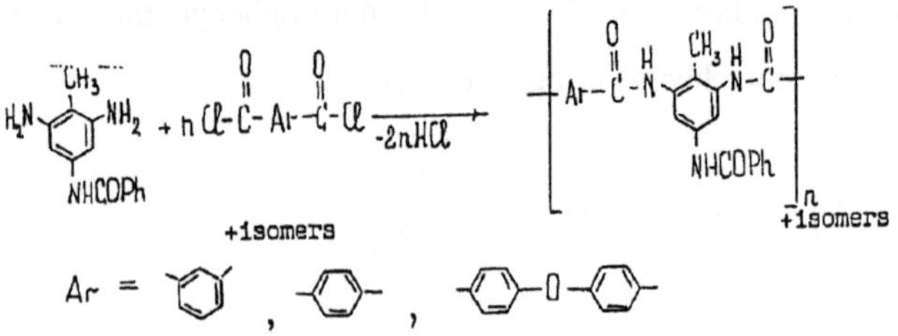

and polyimidation:

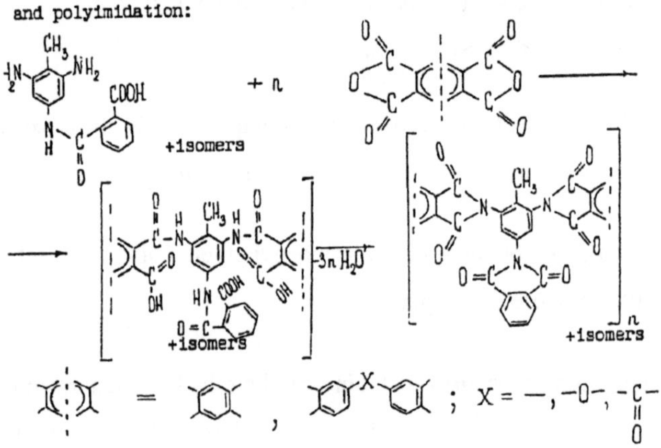

The presence of target polyamides and polyimides of benzamide and N-phthalimide substituents in macromolecules leads to the solubility of these polymers in organic solvents. This permits to solve one of the major problems in chemistry of aromatic condensation polymers - their processability.

Another way to obtain condensation polymers from TNT is the application of RO-substituted 3,5-diaminobenzenes as monomers. The most accessible methoxy- and phenoxy-3,5 diaminobenzenes (MDAB and PDAB, respectively) as condensation polymers were studied in detail.

These substituted m-phenylenediamines were used to obtain polyamides and polyimides with methoxide and phenoxide substituents [11,12].

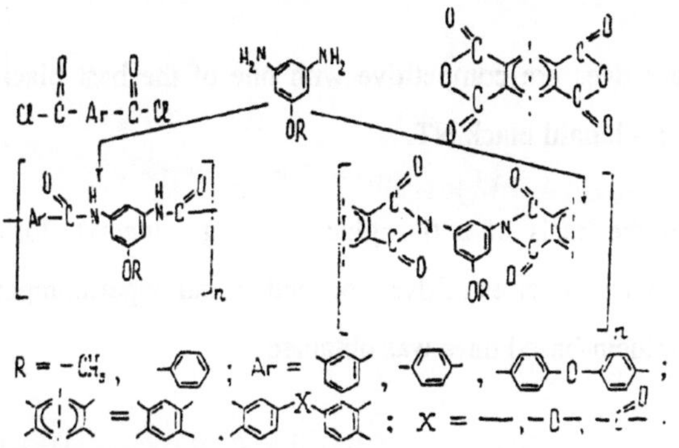

Polyamides and thermostable polyimides thus obtained are soluable in organic solvents. On the basis of some of them firm elastic films were produced.

3. Dyes on the basis of products of trotyl processing

Benzoid tri- and diamines formed in the process of the present work were studied as semiproducts for producing new azo dyes. The main attention was paid to the most accessible substances: 2,4,6-triaminotoluene (TAT), methoxy-3,5-diaminobenzene (MDAB), and phenoxy-3,5-diaminobenzenes (PDAB). The presence of amino groups and the ability of azo couplings to enter reactions as azo components allowed to elaborate a number of schemes for the synthesis of cotton and leather dyes in a wide range of colors. As diazo components we used the products manufactured in Russia on a commercial scale for producing standard dyes.

It is generally known that black and brown dyes are of most demand. Similarly to the known ones, direct black dyes on the basis of TAT, MDAB and PDAB were secured. They are highly competitive in properties with the known dyes and are recommended for commercial manufacture by MNPO „NIOPIK“.

A general formula for two black direkt dyes on the basis of TAT and MDAB is given below.

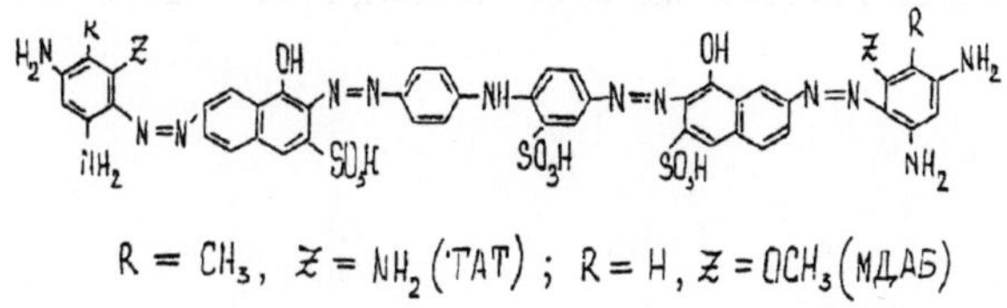

The properties of these dyes are competitive with one of the best black dyes for leather manufactured by BASF - lunalil black NT.

A search for dyes on the basis of a floroglucin analog - methyl floroglucin as an azo component was undertaken. A series of dyes for leather and paper competitive in color and fastness with the floroglucin-based ones was obtained.

4. Conclusion

In the course of the present project on chemical conversion of explosives a scientific base and techniques for TNT processing into condensation monomers, polymers (polyimides and polyamides), and dyes were created.

The elaborated synthetic methods allow to control flexibly and purposefully monomer and semiproduct structures by introducing necessary fragments and thus to impart new valuable properties to the polymers and dyes made on their basis.

References

1. M. Lubben, H. Scholles, K. Reichelt, K.H. Kluger, Proceed. 23rd Int.Conf. of ICT, Karlsruhe, Germany, 1992, pp-8-1 - 8-9.

2. A.R. Mitchell, R.D. Sanner, Proceed. 24th Int. Conf. of ICT, Karlsruhe, Germany, 1993, pp.38-1 - 38-6.

3. A. McKillop, B.D.Howarth, R.J.Kobeylecki, Synthetic Commun., 1974, v.4, p.35 .

4. F. Terrier, Chem. Rev., 1982, v.82, p. 77.

5. G.A.Artamkina, M.P.Egorov, I.F.Beletskaya, Chem. Rev., 1982, v. 82, p. 427.

6. V.N. Solkan, S.A. Shevelev, Izv. RAN, ser. khim., 1993, No11, p.1892.

7. V.Gold, C.H.Rochester, J.Chem.Soc., 1964, p. 1692.

8. P.T. Izzo, J. Org.Chem., 1959, v.24, p. 2026.

9. F. Effenberger, M. Koch, W. Streicher, Chem.Ber., 1991, Bd. 124, S. 163.

10.S.A.Shevelev, M.D.Dutov, A.L.Rusanov, I.A.Vatsadze, O.V.Serushkina, A.M. Andrievsky, „A Method of Producing Aryl and Hetaryl 3,5-Dinitrophenol Ethers", Patent of Russia, 1994.

11. A.L. Rusanov, L.G.Komarova, A.M. Trushkin, S.A. Shevelev, M.D. Dutov, O.V.Serushkina, A.M. Andrievsky, Vysokomolekul.Soed., ser.B, 1993, v.35, p. 883.

12. S.A.Shevelev, M.D.Dutov, A.L.Rusanov, L.G.Komarova, M.P.Prigozhina, A.M. Andrievsky, „Phenoxy-3,5-Diaminobenzene as a Monomer for the Synthesis of Polymers. Polyimides and Polyamides on its Basis as Thermostable Materials with Improved Processability", Patent of Russia, 1993.

General properties of TAT-based N-phthalimido--substituted polyimides of general formula:

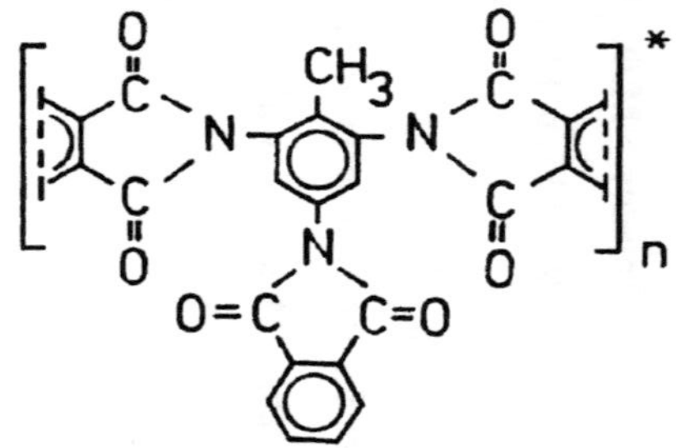

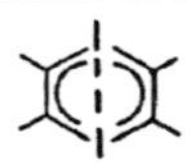	η_{red}, N-MP, 0.5%, 25°C	M	$T_{soft.}$, °C	$T_{10\%}$ DTGA, °C	Solubility			
					N-MP	DMF	$CHCl_3$	TCE: phenol
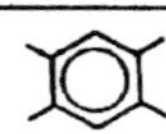	0.30		–		ps**	ins	ins	ins
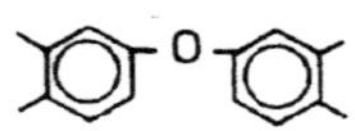	0.30		338	400	s	s	ins	s
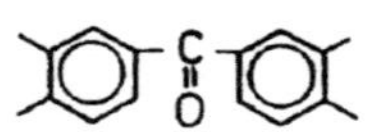	0.30		340	405	s	s	ins	s

* Polyimides obtained on the basis of free TAT base.
** 90% solubility in N-MP

General properties of the polyimides of general formula:

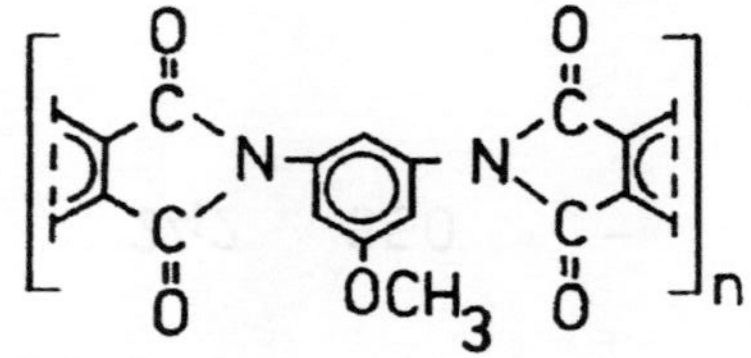

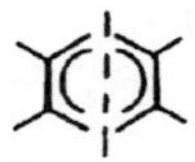	$\eta_{red.}$, 0.5%, 25°C dl/g	$T_{soft.}$, °C	$T_{10\%}$ DTGA (air, T=45° min)	Solubility			Films properties, 25°C	
				N-MP	m-cresol	TCE: phenol	σ, MPa	ε, %
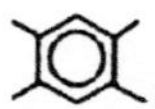	1.12* —**	—	400	ins	ins	ins	—	—
	0.83* —**	310	400	ins	ins	ins	—	—
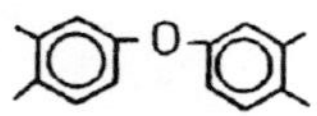	0.89* 0.82** 0.47***	300	440	ins	s	s	1140	9

* Viscosities of (poly-o-carboxy) amides solutions MP.
** Viscosities of polyimides, obtained by catalytic polycyclization of (poly-o-carboxy-amides) using complex pyridine-acetic anhydride (TCE:phenol)
*** Viscosities of polyimides, obtained in m-cresol

Properties of the Polyimides of General Formula:

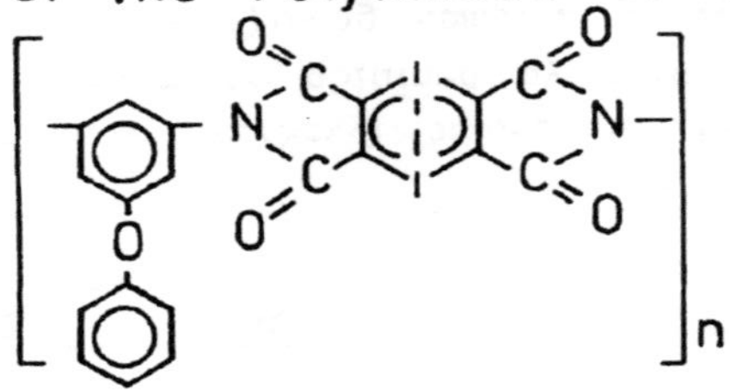

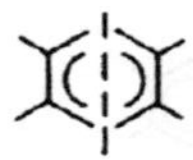	React. med	$\eta_{red.}$, dl/g	M_w	Tg, °C	$T_{10\%}$, °C	Solubility			Films properties, 25°C	
						N-MP	MC	TCE-p	σ, MPa	ε, %
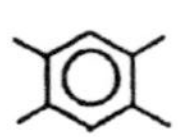	N-MP	0.66	–	380	550	ins	ins	ins	–	–
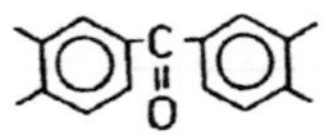	N-MP	0.44 0.35	15000	260	530	s	s	s	brittle	film
–〃–	MC	0.69	–	275	–	s	s	s		
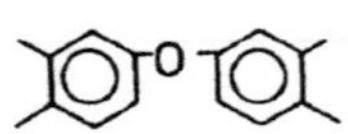	N-MP	0.83	48000	245	540	s	s	s	1080	8
–〃–	MC	0.94	–	242	550	–	–	–	–	–

Some properties of the polyamides of general formula:

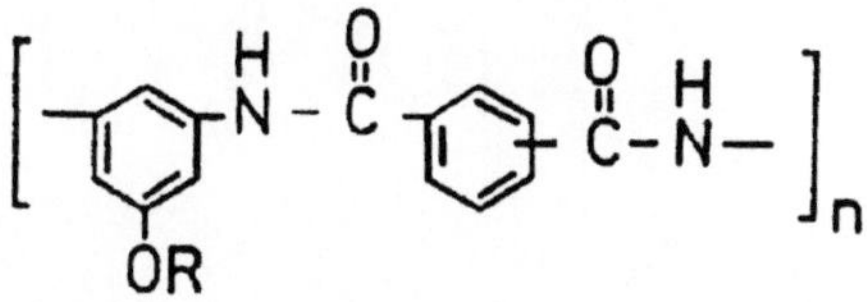

-R	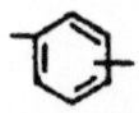	η red., N-MP, 25°C dl/g	mol. weight+	Tg, °C	T10%, TGA, °C	Yield, %
$-CH_3$	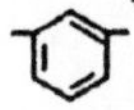 *	0.25	9500	224	330	100
$-CH_3$	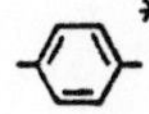 *	0.50	21000	256	370	96
$-C_6H_5$	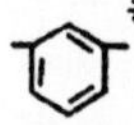 *	0.55	23000	262	356	100
$-C_6H_5$	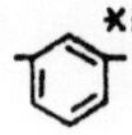 **	0.46	31000	246	400	98
$-C_6H_5$	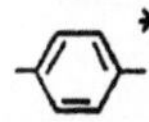 *	1.05	58000	275	400	100
$-C_6H_5$	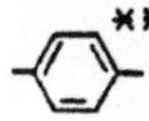 **	0.38	17000	236	350	95

* Polyamides synthesized in N-MP
** Polyamides synthesized in chloroform with triethylamine

RE-USE OF 155 MM HE AUXILIARY PROJECTILES FOR PRACTICE ROUNDS AND INDUSTRIAL EXPLOSIVES

ING. P. DE WIT
Naval Weapon Establishment
Ammunition Department
Royal Netherlands Navy
Den Helder

Abstract

The process of ensuring a well-trained and ever-ready military force nowadays also involves ensuring that the exercise grounds or sites suffer as little as possible from adverse influences such as noise, pollution and extra hazards - such as unexploded shells.

The fact that adjoining built-up areas, residential areas, villages and towns also suffer from the effects of military exercises is a fact that has to be given serious consideration now that the public views such actives much more critically, and legal rules and regulations are becoming stricter.

All these factors emphasise the necessity of producing ammunition which, after serving its purpose for combat use, can be converted to environmental-friendly practice ammunition.

Production process for the inertisation of grenades

H. Krause (ed.),
Conversion Concepts for Commercial Applications and Disposal Technologies of Energetic Systems, 151–162.

Preface

What is the Ammunition Department and what are its chiefs tasks?

The Ammunition Department is a main division within the Royal Netherlands Navy's Weapon Establishment and is certified by AQAP-1(ed3)ISO-9001. This certificate issued under number 088-10 is valid till December 11th 1994. Its purpose is: to maintain the RNLN's ammunition as economically as possible using the means at its disposal.

Its duties include:
- assembling, servicing and overhauling conventional ammunition and detonators;
- testing, servicing and repairing guided missiles and torpedoes;
- carrying out surface treatments and repairing shells;
- disposal off ammunition;
- maintaining a safety organisation which ensures that the safety regulations applying to this establishment are strictly observed.

Introduction

Producing this practice ammunition economically, modifying it and servicing it, require extra care not only because it is "explosive", but also because production has to be efficient providing the high "output" and "quality" required.

In developing such a process, no matter how automated, the human element is still an indispensable link. The ultimate procedure therefore automatically needs to incorporate all aspects of ergonomics, safety and regulations governing the working environment.

The four main aspects are:
- explosion hazard;
- environmental friedliness;
- ergonomics;
- environment regulations applying to working conditions.

These four factors will have a significant effect of determining the production process.

Also as an added dimension extra attention will need to be paid to the entire process from the point of view of the working environment because explosives used in ammunition are always of a toxic nature.

Process designing

Splitting out the process of designing, the 155 mm production line gives three main objectives namely:

a. determination of <u>all</u> problems referring to the four main aspects mentioned on page 4;
b. removing the explosive charges;
c. inertation of the grenades.

A. THE FIRST MAIN OBJECTIVE is to determine the problems referred to here and take them into account in the design phase of the production line.

In other words, adopt preventive measures.

Here, as everywhere, prevention is better than cure!

When setting up the production process for 155 mm shells for the Royal Netherlands Army, which have been serviced in the Royal Netherlands Navy's Ammunition Department since December 1991, special attention was paid to the aspects mentioned previously, i.e. preventive measures. Once these have been incorporated into the overallprocess, fewer problems will arise.

B. THE SECOND MAIN OBJECTIVE, with the above aspects in mind, is how to remove 6,5 kg of TNT and adapt the shell so that, after inertiation during practice it retains the same ballistic characteristics as the live shell and indicates that it works properly also at low temperatures.

Up to now in the RNLN two production methods have been used for this purpose, namely bore out or melt down (see fig. 1).

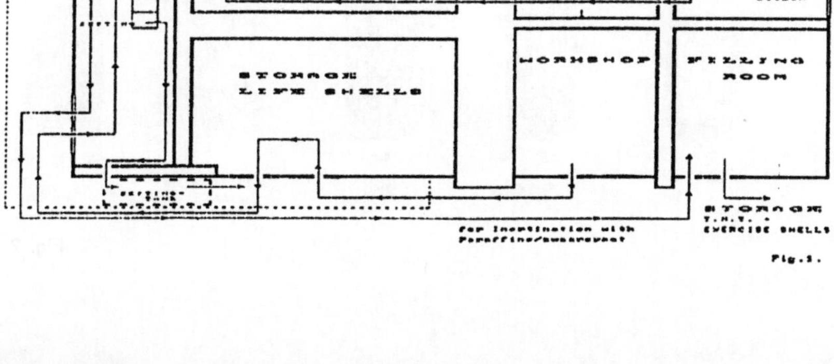

Fig.1.

Method 1 can be used only when the inside of the shell is perfectly cylindrical in shape (for example the 76/62 oto melara ammunition) but this is often not so.

Method 2 melting down requires a tremendous amount of energy as the entire shell, 42 kg in weight, has to be heated to a minimum of 80 degrees Celsius and heating 36 kg of steel results in a higher energy loss. Also, for safety reasons, the temperature of the steam must never exceed 100 degrees Celsius as explosives can start to decompose at 150 degrees Celsius. This must never be allowed to happen (explosion hazard).

The Royal Netherlands Navy's experience with these processes was far from satisfactory where the level of output and impact on the environment were concerned. But it was mainly the efficiency of the production process that prompted the decision to look for and entirely new procedure.

Also the concentration of TNT (as vapour) in steam is many times greater than when using a cold water jet system and finally after much careful consideration, it was decided to obtain for scouring with a water jet installation at a pressure of 1100 bar (see fig. 2 and 3).

The criteria considered when making this final choice were the four aspects mentioned previously and the considerable improvement in output from the production process.

In accordance with this choice we had to ask our selves what the biggest problems were whilst removing explosives by means of water under high pressure.

These were the amount of water and the pollution with explosives and also the conveyance of energy collected in the water as heat. Also, the water explosive vapour which is generated by the process, because of the extreme high pressure and reverse of energy, must be recovered without any impact on the environment.

To do so there was only one good solution, "bring back everything in the process itself".

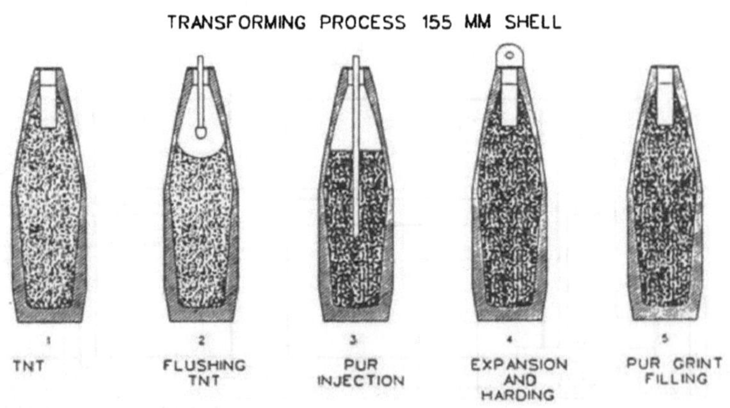

Fig. 2.

Clean the water and return it to the supply tank of the pump installation. Condense the vapour from the ventilation systems and transport the water with explosives also to the supply tank of the pump installation.
As a result from previous mentioned items we have to do a big improvement of the working environment in the explosive area, points of considerations are:
1. a better ventilation system;
2. absolute separation between explosives and non-explosives area's (see fig. 3);
3. improvements in the process it self;
4. reclaiming explosives out off the ventilation system.

After implementation of the various improvements this proved to be the success formula and in practice I can say, that we are running the process with the same amount of water from the last two years and producing 38000 exercise grenades.

Till now it saves us a lot of money, because of not using approximately 240 m^3 of water a day, being the 96 lts/min capacity of the pump installation, and furthermore there is no impact of the environment and recent measurements carried out by TNO and our marine Chemical Laboratory conformed our expectations.
There was no pollution in any way or danger to:
1. workconditions;
2. environmental;
3. safety;
4. ventilation air.
Everything was in good condition.

PROCESS ROUTING: 155 MM SHELL TO NR.129 SHELL

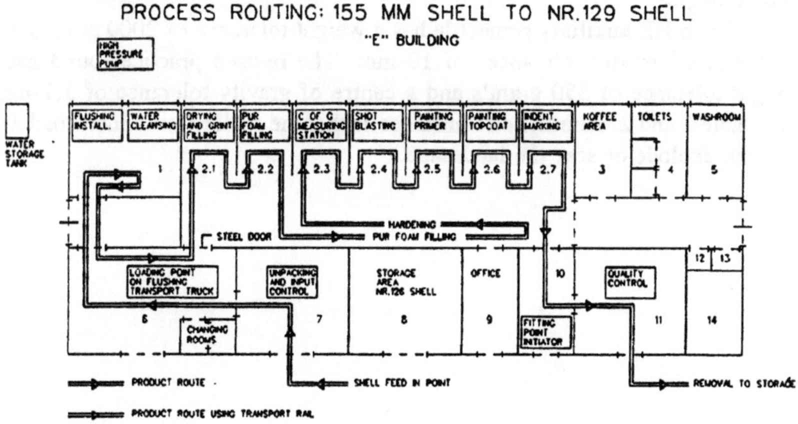

Fig. 3.

These improvements gave the people who are working in the explosive area much more satisfaction and off course motivation to carried out there task, stay healthy and producing every month the expected output with pleasure.

But one big problem remained, namely: "How do we dispose of approximately 450.000 kg of TNT?".

As a consequence of the process we produce TNT with nearly 15 % of water and this composition was very difficult to sell.

Finally we chose a firm which is a specialist in slurry explosives for open pit mining and UTC in Dallas USA, working also for the Ministry of Defence, has proved to be a good partner.

But this is not the final solution for the Ammunition Department of the Royal Dutch Navy.

Together with TNO we are working on a process of burning organic explosives, such as TNT, and using the energy for heating purposes of our total production plant.

Than the circle is round and we are using all relevant products out off the conversion from the 155 mm shell our selves and do not harm the environment in any way.

C. THE THIRD MAIN OBJECTIVE, is inertisation of the grenades by means of a contents that have the same centre of gravity and weight as the life projectiles. The inert filling is off a special composition, which can be used within a temperaturerange of - 40 degrees Celsius to + 80 degrees Celsius and is ballistically matched with life rounds. The main difference is a 33 % reduced safety template because of his accuracy in weight and centre of gravity.

For example:

A 155 mm HE auxiliary projectile has a weight tolerance of 2000 gram's and a centre of gravity tolerances of 10 mm. The re-used practice round has a wight tolerance of 350 gram's and a centre of gravity tolerance of 3,7 mm, see table 1 and 2. Another circumstance is that the shells have a inert load and do not explode or scar the landscape.

A.

TABLE 1: specification 155 mm HE project M107

* Mass range				
Range	Filled shell (exclusive casing)			
	Kilogram		Mark	
	from	up to and		
2	40.823	41.413	X X	
3	41.322	41.912	X X X	
4	41.731	42.910	X X X X	-- standard
5	42.320	42.910	X X X X X	

B.

TABLE 2: specification 155 mm HE project M107

* Centre of gravity		
Range	Filled shell (exclusive casing)	
	Millimetres	
	from	up to and
2 to 5	230	250

* CofG measured from the bottomside of the Grenade.

C. Results of a half day production, semienumber 26 date 03-25-1994.

TABLE 2: Results 155 mm practice projectnumber 129					
Centre of gravity (CofG) measurement 155 mm mass range					
Nr.	Weight	CofG	Deviation CofG	Deviation tolerance	time
1	42.180	241.0	+1.0		07:39
2	42.093	241.5	+1.5		07:45
3	41.931	239.5	-0.5		07:51
4	41.942	239.5	-0.5		07:57
5	42.013	239.6	-0.4		08:02
6	42.166	241.2	+1.2		08:08
7	41.997	240.2	+0.2		08:13
8	42.159	243.3	+2.3		08:18
9	42.150	241.2	+1.2		08:24
10	41.988	239.5	-0.5		08:28
11	42.013	240.4	+0.4		08:32
12	41.966	241.9	+1.9		08:35
13	42.229	241.5	+1.5		08:38
14	42.000	239.8	-0.2		08:41
15	41.922	240.9	+0.9		08:45
16	41.967	240.5	+0.5		08:49
17	41.878	239.9	-0.1		08:52
18	41.886	238.9	-1.1		08:55
19	42.000	240.4	+0.4		08:58
20	42.006	240.4	+0.4		09:01
21	41.992	240.1	+0.1		09:04
22	41.901	240.5	+0.5		09:07
23	41.905	241.0	+1.0		09:09
24	42.134	240.1	+0.1		09:12
25	41.883	240.1	+0.1		09:15
26	41.842	239.7	-0.3		09:20
27	41.887	239.4	-0.6		09:25
28	41.888	240.7	+0.7		09:29
29	41.892	239.4	-0.6		09:40
30	41.865	238.6	-1.4		09:45
31	41.904	239.3	-0.7		09:51
32	41.881	241.0	+1.0		09:56
33	41.874	240.3	+0.3		10:00
34	42.108	240.5	+0.5		10:30
35	42.055	240.3	+0.3		10:33
36	41.997	240.3	+0.3		10:37
37	41.980	240.0	+0.0		10:40
38	42.134	240.5	+0.5		10:45
39	42.136	241.3	+1.3		10:48
40	41.987	239.4	-0.6		10:51
41	42.001	240.4	+0.4		10:54
42	42.056	240.6	+0.6		10:57
43	41.860	240.2	+0.2		10:59
44	41.890	238.6	-1.4		11:03
45	41.917	239.8	-0.2		11:10
46	42.207	242.5	+2.5		11:14
47	42.207	242.5	+2.5		11:15
48	42.068	240.9	+0.9		11:19
49	41.988	242.1	+2.1		12:19
50	41.920	241.3	+1.3		12:21

A product from start to finish

life and inert ammunition

D.

TABLE 3: Results 155 mm practice projectnumber 129

* Centre of gravity

Range	results all grenades in range 4		
	max.	min.	average tolerance of this serienr.
Weight/ gram	42.229	41.860	± 0,369
centre of gravity in mm	242,5	238,6	± 3,9

Remarks

After refurbishment we can say that all grenades fitting in range 4 off the mass range's and that is a very big improvement were it concerned centre of gravity and weight.

To keep track of a big production output, especially in the range of the above mentioned details, an automatic 100 % control system is installed in the production line. This computer system collects all the information and gives a warning when the grenade does not fit in his tolerances. This particular grenade is re-used in the process it self. Together with the workorder a print-out is delivered, for the complete series off 1400 grenades each, to the customers as an addition to his calculation or shooting results (see table 2).

A surface treatment assembling and packaging completes the whole process of refurbishment.

Enclosure

MINISTERIE VAN DEFENSIE
DIRECTORAAT-GENERAAL
MATERIEEL
Nederland

MINISTRY OF DEFENCE
DIRECTORATE-GENERAL
OF MATERIEL
The Netherlands

KWALITEITSZORGCERTIFICAAT

QUALITY ASSURANCE CERTIFICATE

Hierbij wordt verklaard dat het kwaliteitssysteem van:
This is to certify that the quality system of:

BEWAPENINGSWERKPLAATSEN DER MARINE

voldoet aan de voorwaarden gesteld in:
meets the requirements specified in:

AQAP - 1 (ed. 3) / ISO - 9001 (1987)

voor onderstaande produkten, bewerkingen en/of diensten:
for undermentioned products, processes and/or services:

Installation, integration, adjustment, setting to work and testing of weapon systems. Repair, overhaul, modification and renovation of weapon systems, fire control systems, guns, launchers, guided missiles, ammunition, torpedoes, fire extinguishers. Technical support and engineering services such as supply support, technical documentation, configuration management, calibration, data collection and analysis, engineering investigations and consultancy.

Plaats(en) van fabricage: Den Helder
Plant(s):

Dit certificaat uitgegeven onder nummer: 088-10
This certificate issued under number:

is geldig tot: 11th December 1994
is valid till:

DE DIRECTEUR-GENERAAL MATERIEEL

Mr J. Fledderus

's-Gravenhage, 11th December 1992

References

1. Machacek, Oldrichs and Eck, Cary R. (1992) *Waste propellant and smokeless powders as ingredients in commercial explosives*, Conference of ICT Karlsruhe FRG.
   Clarck, Ron P. (1992) *Waste propellant and smokeless powders as ingredients in commercial explosives*, United Technolics, chemical System Division, San Jose California.

2. Bartelds, ir H and Temmink, ir. H.G. (1992) *Feasability study environmentally acceptable disposal of ammunition and explosives d.d. May 1992*, TNO/IMET, dossiernr. 112326-222253.

3. TNO, (1993) *TNT vapour/Dust concentration measuring during 155 mm refurbishing process, reference nr. PML 1993-6.*

4. TNO, (1989) *PML research removing TNT from waste water by active coal*, reference nr. PML 1989-23.

5. *Ammunition Department production process, Quality Accepting Plan 155 mm* (1991), document nr. KIP OS BG nr. 129C, OBG153 155 mm HW.

6. *Dutch regulations about storing, transportation of ammunition and explosives* (1992) MP40-20.

7. Naval Sea Systems Command (1993) *Ammunition and Explosives ashore, safety regulations for handling, production renovation and shipping*, Washington DC, NAVSEA OP 5 Volume I

USAGE OF EXPLOSIVE TECHNOLOGIES FOR NATIONAL ECONOMY

ASSOCIATE MEMBER OF THE RUSSIAN ACADEMY OF SCIENCES,
CNIIHM DIRECTOR, YU. I. KRASNOSHCHYOKOV

High-skilled scientific and engineering personnel has been freeing as a result of conversion in ammunition branch of Russian defence industry. Their experience in explosive transformations makes it possible for explosion, being a typical military phenomenon, to come to peaceful service.

For these purposes scientists of CSRICM altered this process significantly, having turned its nature from destructive into creative.

It allowed to come to new technological findings, featuring uniqueness and enabling to achieve new results.

To illustrate this idea I shall give some examples of new engineering developments. Let me express my hope that solving such problems is of interest for you. My survey will be brief enough because detailed information can be given to you directly at the exhibition.

1. Explosive (impact) division of high-volume metal constructions

This development is aimed at fragmenting old ships, barges, planes and so on to make scrap of them.

The essence of the development is to introduce into the process an operation of instant explosive disintegration of the construction into large fragments (for example, decks, boards, upper constructions), which makes the following process of traditional processing easier and quickens it too.

H. Krause (ed.),
Conversion Concepts for Commercial Applications and Disposal Technologies of Energetic Systems, 163–167.

So two problems to been solved:

- time and cost saving;
- improvement of working conditions, including personnel protection of harmful influences.

Possible co-operation:

- services, rendered by Russian teams;
- creation of joint-ventures for scrap-making

2. Preventive facility, making roads, health-resorts and settlements avalanche-safe

The scope of the problem is great enough. By their impact action powerful avalanches are comparable with a nuclear explosion, having the TNT equivalent of a few kilotons. Avalanches are observed in more than 80 countries. Sums of damage are estimated 1...1.5 billion dollars.

The method is worked out on the basis of fundamental study of avalanche formation and go-down.

The main feature of the method, making difference between it and traditional approaches, is the possibility of relieving „immature“ avalanches.

It is very important because it enables to make all the avalanches zones safe enough. The use of helicopters instead of artillery guns provides high mobility.

Possible co-operation

- services, rendered by Russian teams
- creation of joint-ventures to render these services
- sales of completed facilities for protection or influence elements.

In the last case the reconstruction is needed to make it possible to use western carriers.

3. Explosive manufacture of superhard materials such as boron nitride and polycrystalline diamond

As a result of fundamental studies a process using dynamic explosive high-pressure loading followed by static holding at the pressure of a few tens of kilobars which enables to perform phase transformation of hexagonal boron nitride into a diamond-like modification of boron nitride has been developed. Wear-resistance of the tool (the drawing die) increases 7...10 times to compare with hard-alloy materials.

Possible co-operation is to create a joint-venture to manufacture and realize superhard materials and articles of them. Investments are needed.

4. Explosive surface-hardening of metal parts

The essence of the method is in precision adjustable application and attachment of thin explosive along the contour of a part surface to be processed. It enables to achieve graduated explosive action onto the surface. Figures of hardening is given in the figure.

Possible co-operation is joint taking out patents for the technology followed by licence selling.

5. Explosive removal of cutting tool debris

A special liquid explosive is invented for this technology. Moscow Automobile Plant ZIL has accumulated experience of using the method.

Possible co-operation is joint taking out patents for the technology followed by licence selling.

6. Automatic system of emergency prevention in the cases of gas and dust explosions in mines

Similar systems have been developed in serveral countries (in France, Germany, Holland).

Our development has the following advantageous features:
compactness of the sample, safety for personnel when working, and mobility.

Function:
A sensor of the system founds the nucleation site of the explosion and gives the command to fire the powder inhibitor, which confines the explosives spreading. The system enables multiple usage.

Possible co-operation is offered to create a joint-venture to finish development, qualify the system and to manufacture it in Russia.

7. Cultivator for sown areas designed to destruct soil crust (primarily for cotton plantations)

The problem: agricultural fields in hot-climate regions (Central Asia, Middle East, Australia, Southern America) suffer surface crust formation as a result of quick soil drying after rains. The thickness of the crust reaches 40 mm. Under such crust seeds cannot germinate.
Traditional mechanical tools like harrows are not effective, because they injure seeds and sprouts. In Uzbekistan they have to re-sow up to 400,000 hectares of cotton.
So the problem was to find a way to destruct the crust without damages to seeds and plants.
The solution of the problem was almost paradoxical: to use an explosion for this purpose.
The essence of the method is to destruct the crust by gaseous mix explosion in a tube, shock action being spread along the crust surface only. Not to frighten prospective customers I must say that the system is absolutely safe. In spite of explosive nature of the action any

given explosion uses only 1 g of petrol-air mix (0.1 g of petrol). As a matter of fact we deal with a micro-explosion.

Besides its main effect (soil-crust destruction) the process also improves harvests (data in the figure).

Possible co-operation is to create a joint-venture to manufacture and to realize cultivators or to sell licences. Joint studies aimed at widening fields of application of the method are also possible.

8. Intensification of oil and gas output

It is known, that in the process of well exploitation as a result of deposition of paraffin, asphalt-resin and other substances at the areas of oil intake, the volume of oil flowing from the well decreases. Some ways „to revive“ the well are known too. But all of them are very expressive.

According to the development the lower part of the well is under the action of combustion products of special powder charges. It results in creation of a high-pressure zone which is followed by pulsation with decreasing amplitude.

Because of these factors at near-well zone residual fractures are created, which provide good hydraulic connections between the well and remote zones of the bed. At the same time the near-well zone is cleaned of asphalt-resin substances, paraffin and chemical reaction products.

Possible co-operation is to create a joint-venture to carry out these works, to sell licences and prototypes and go on with the development of the method.

INDUSTRIAL EXPLOSIVES BASED ON DOUBLE-BASE PROPELLANTS

B.M. Balojan, V.V. Galkin, D.F. Marov,
N.I. Konev, V.V. Falko, Russia

Application of the double-base propellants (artillery and rocket ones) as industrial explosives is reported. The main scientific and technical criteria of development of industrial explosives for open mining (granular powders), detonator charges for seismic prospecting, detonator charges for blasting of wells and also linear shaped charges for crushing large metal structures and outsize rocks are discussed.

Directions of the development of the industrial technology for production of above explosives are also given.

This paper deals with the scientific study and development of a new class of industrial explosives on the basis of recovered double-base propellants (artillery powders: DG-2, DG-3, NDT; rocket propellants: RSI, NMF) and also with development of industrial technologies for their production with maximum application of the present apparatus and technologies.

This problem depends on the demilitarization and conversion of the military industry /1/.

There are some general considerations:

- The powder due to its nature is a powerful source of concentrated energy. This energy can be realized in two regimes:
- detonation, when explosion results in a work because of instantaneous generation of a great amount of gases,
- controlled burning, when generation of the gases results in regulated work.

H. Krause (ed.),
Conversion Concepts for Commercial Applications and Disposal Technologies of Energetic Systems, 169–182.

1. In both cases all processes are programmed.
2. Powders have properties (density, water-resistance), that exceed the properties of the industrial explosives; however their sensitivity to mechanical stress and detonation waves is inferior.
3. Recovered powders depending on the removable from the service ammunition have different compositions, shape-dimensions and properties. So the differentiated ways of their processing into industrial explosives are needed.
4. Profitable production is the main aspect of the development of the new industrial explosives for the national economy.

Taking into account the chemical compositions, dimensions and shapes of the shapes of the recovered powders, it was concluded that thedevelopment of new explosives and technologies is being expedient.

Industrial explosives for the open pit mining

Calculations and experimental research have shown /2/ that all grades of the artillery ballistic-type propellants, rocket nitroglycerine (nitrate) propellants and their mixtures with pyroxylin powders can be a base for the new waterproof industrial explosives. However in order to upgrade the oxygen balance and environment ecology after explosion, ammonium nitrate was studied for application in the new compositions.

Detonation capacity of the compositions vs various pyroxylin powder and ammonium nitrate contents has been studied. / Fig. 1 /

Fig. 2 shows that with increasing content of the pyroxylin powder the more detonation capacity of the mixtures also increases. Compositions containing 15 - 35 % of ammonium nitrate have the highest detonation capacity. The amount of toxic gases released in detonation of the charges is 1.2 to 1.3 times less than that of the Grammonit 30/70 used in blasting.

It is not a good idea to increase ammonium nitrate content in the mixtures for two reasons:

- increasing of the critical diameter of detonation to a great extent,
- decreasing of the length of the bore blasting charge under flooded conditions because of solution of ammonium nitrate.

Fig. 3 shows the detonation rate of the compositions in the dry and in the water-saturated conditions vs the charge diameter.

Analysis and comparison of the above results with the publicated data have shown that the double-base propellants have processing and operation characteristics (safety and ecological purity), which are better than that of the industrial explosives used for blasting.

Joint research work of the OIHF of the RAN, LNPO „Sojuz", the „Russia" chemical plant resulted in development of waterproof granular explosives (Granipory) /4/.
Physico-chemical and explosion characteristics of the Granipory are given in Table 1.

Table 1

Characteristics	Granipory	Granulotol
a. Calculated		
Heat of explosion, Kkal/kg	830 to 850	870
Gas volume, 1/kg	870 to 890	750
Oxygen balance, %	-(45 to 50)	-74
Volume of toxic gases in terms of CO, 1/kg	260 to 280	320 to 330
b. Experimental		
Bulk density, g/cm	0.8 to 0.9	0.9 to 1.0
Critical diameter of charge detonation, mm		
- dry condition	100 to 120	80 to 100
- water-saturated condition	80 to 100	
Detonation rate, km/s		
- dry condition	4.8 to 5.2	4.0 to 4.6
- water-saturated condition	6.1 to 6.3	5.5 to 6.0
Impact sensitivity		
- explosion rate, %	20 to 24	8 to 12
Friction sensitivity, kgs/cm	2180	2500

Resistance of the Granipory to acid and alkaline media is higher than that of the Granulotol. So they are promising for breaking mining and sulphide. Tests of the Granipories have shown their compliance with the requirements to the explosives.

Table 2 shows the main parameters of the explosives and the results of two tests.

Table 2

Explosive	Granipor 3	Gramonit 30 / 70	Granipor 1
	Granite 16	Granite 16	Gabrodiabas
Height of shoulder, m			13 to 16
Dept of well, m	18	18	17 to 20
Diameter of well	216	216	269
Height of water column, m	3 to 8	3 to 8	up to 5
LSPP of the first line, m	7	7	7 to 8
Number of lines	5	5	5
Distance between lines, m	5	5	5,5
Distance between wells, m			
Number of blasted wells	5	63	22
Charge mass in a well, kg	390	510	550
Height of a charge in a well, m	14	6	12 to 13
Consumption of explosive per 1 blasting, kg	22230	30000	13380
Length of drilled well, m	1026	1134	440
Blasting sheme	wedge - cutter line		
Volume of blasted rock mass, m	26800	32000	14300
Yield of rock mass per 1 m of well, m	26.1	28	32.5
Specific consumption of explosive, kg/m	0.82	0.93	0.93
Yield of outsize materials, %	3 to 4	5 to 6	5 to 7
Treatment of shoulder base	good	good	good

The above data show that the main parameters of the blasting (specific consumption, yield of outsize materials) when the Granipories are used are better than that of the Gramonite 30/70.

Six grades of the Granipories and two grades of the water-contained explosives have been developed on the basis of the granular powders and of ammonium nitrate. All these explosives are patented.

To modify some Granipory properties (oxygen balance, sensitivity to mechanical action, ecological effect of blasting) /5/, ammonium nitrate was introduced in the composition of these powders.

The technological scheme for the manufacturing of the industrial explosives based on the recovered powders with ammonium nitrate is given in Fig. 4.

Technological and economical estimations show the competitiveness of the explosives, if the in Fig. 4 shown diagram is used and the total volume of production exceeds 1500 t per year.

Ballistite linear shaped charges (BLSC)

Cutting of huge metal structures and crushing outsize rocks is of great importance for the national economy. Gas-arc cutting of metal structures is very labour-consuming and ecologically hazardous and does not provide cutting of the large reinforced concrete and metal structures. The shaped cords and extended shaped charges are expensive and are not efficient for cutting of the complex structures, have limited diameters and don't have a production base. /6,7,8,9,10/

The object of this work is:

- development of the ballistite shaped charges with the diameters from 10 to 100 mm and good strain properties for a wide temperature range /+ 50 deg C/;
- development of the industrial high productive continuous technology for BLSC manufacturing, including usage of the metal-polymer with maximum application of the existing processes and apparatus for manufacturing of the ballistite propellants;
- improvment of the BLSC operation properties.

Technical Characteristics of BLSC and their compositions are given in Table 3.

Fig. 5 shows their efficiency. Table 4 gives the dimensions of BLSC.

Fig. 6 shows ways of their initiation. The technology of BLSC production is shown in Fig. 7. Comparative characteristics of the linear charges are given in Table 5.

Conclusion

The R & D to substantiate the ways of creation of the new industrial explosives based on the recovered double-base propellants are discribed.

Waterproof industrial explosives for the open mining have been developed. Tests have shown their efficiency under varied conditions of rock breaking. Their operation characteristics are the same as that of the existing explosives and their water-proofness is better. A number of the new linear shaped charges have been developed for cutting of huge metal structures and outsize rocks. The tests have shown their high efficiency.

The Production-technology of the granular industrial explosives and linear shaped charges has been developed. The Techno-economical calculations have shown the charges to be profitable.

References

1. P. Zhukov et al. „Future usage of ballistite powders in national economy". Collection „The main directions of usage of the rocket propellants in the national economy", ed. by Academician B. P. Zhukov, Moscow, 1990.
2. A. Nishpal „Receptivity of the ballistite solid propellants to the detonation when they are used as explosives." Coll. „The main directions of usage of the rocket propellants in the national economy", ed. by Academician B. P. Zhukov, Moscow, 1994.
3. P. Pozdnjakov, B. D. Ross, Reference book on the industrial explosives and esplosion devices. M. Nedra, 1977.
4. V. Galkin, B. M. Balojan et al. „ New waterproof industrial explosives based on the solid propellants and powders", Gorny zhurnal, N6, pp37-39, 1992.
5. M. Balojan, V. V. Falko et al. „Usage of the ballistite propellants as industrial explosives",. Coll. ed. by Academician B. P. Zhukov, Moscow, 1990.
6. TU 84. 520-296-79. Cord shaped charges.
7. Catalogue of European Pyrotechnic Cevices for use in space (ESA-Sp 1021), European Space Agency.
8. Patent 75537223 (France).
9. Patent 3185089 (USA).
10. Patent 3782284 (USA).

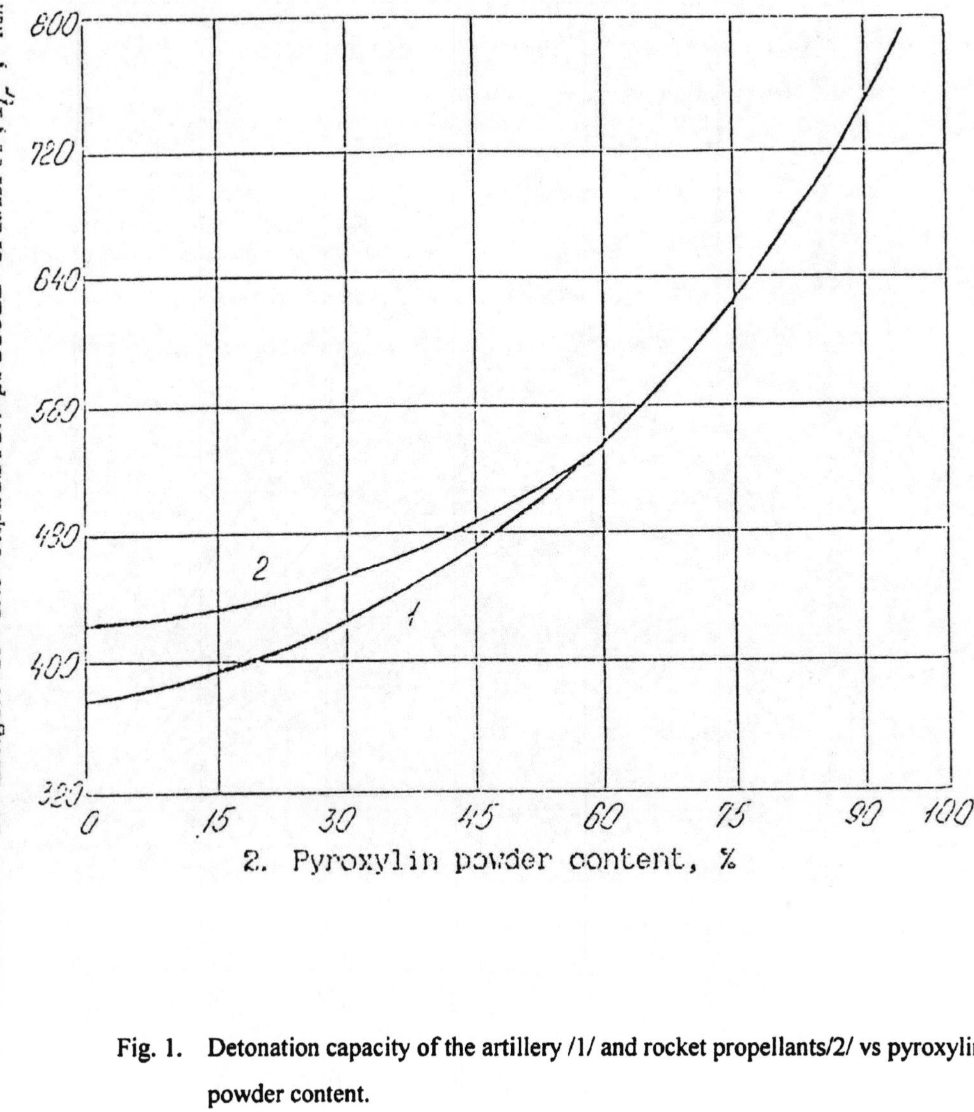

Fig. 1. Detonation capacity of the artillery /1/ and rocket propellants/2/ vs pyroxylin powder content.

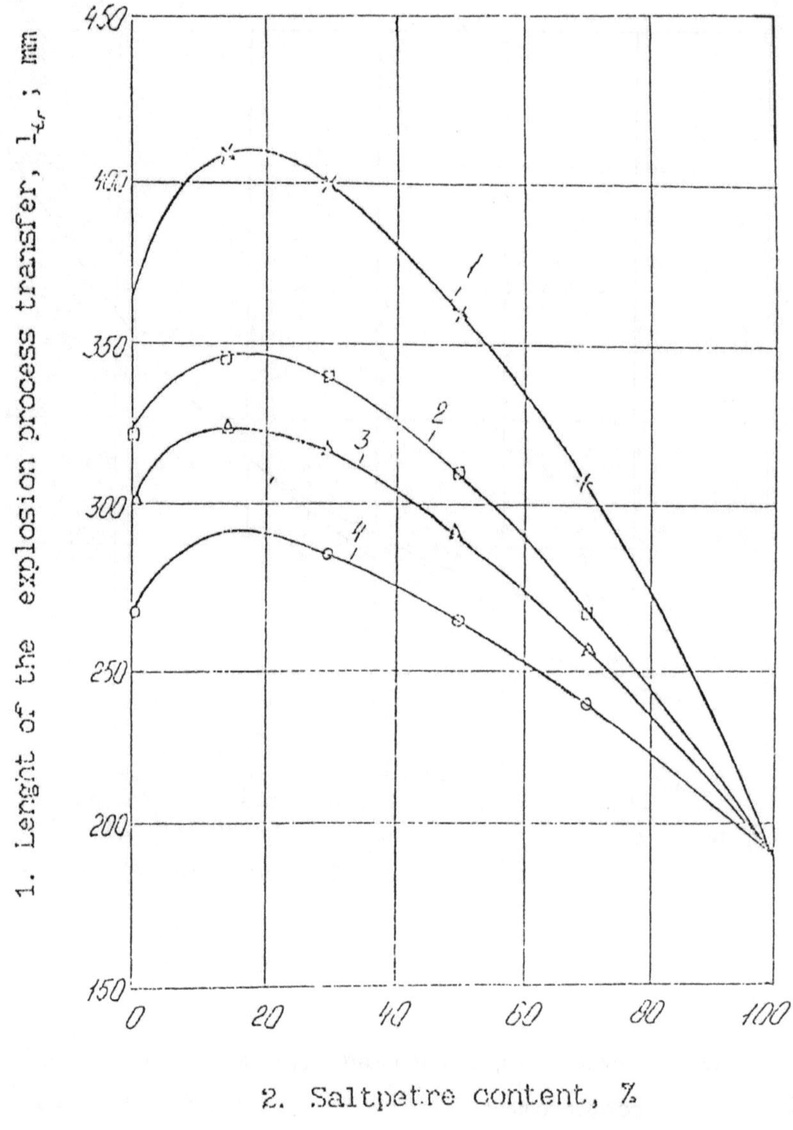

Fig. 2 Diagramm of the detonation rate vs ammonium nitrate content.

1-artillery propellant, 2-artillery propellant + pyroxylin powder

3-artillery propellant + stabilizer,

4-artillery propellant + pyroxylin powder + stabilizer.

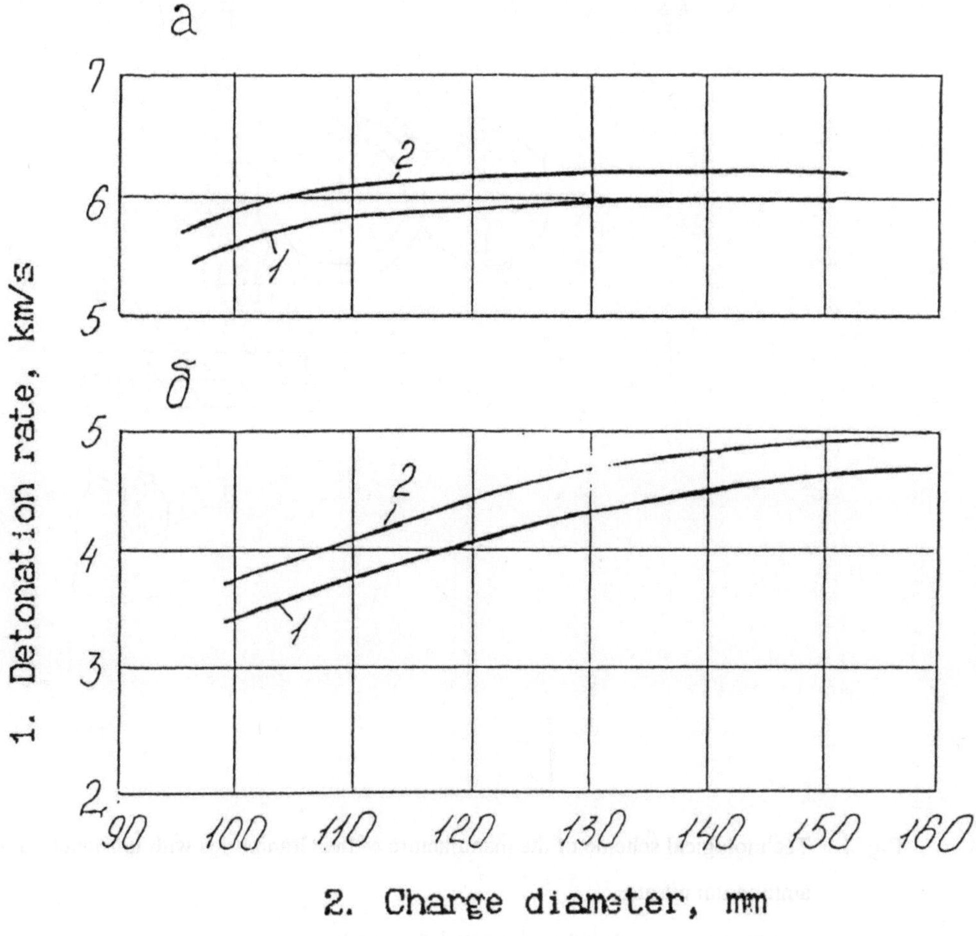

Fig. 3 Diagram of the detonation rate of the artillery üropellant + pyroxylin powder /1/ and rocket propellant + pyroxylin powder vs the charge diameter in the dry /b/ and watersaturated /a/ conditions.

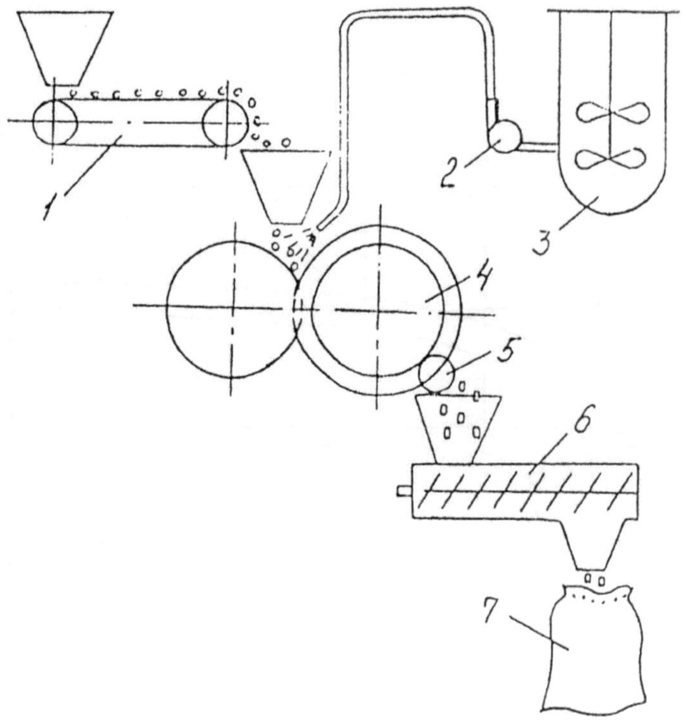

Fig. 4 Technological scheme of the manufacture of the Granipories with introduction of ammonium nitrate.

1-conveyer for charging of the powder; 2-pump;

3-tank for saltpetre solution; solution; 4-rollers-granulator;

5-blade; 6-screw-conveyer; 7-discharge container.

Table 3

Technical characteristics of BLSC

Charge diameter, mm	10 to 100
Thickness of cutted	
abstacle (steel 3), mm	7 to 80
Density of explosive, g/cm	1.65 to 1.7
Trotyl equivalent	0.9 to 1.1
Detonation rate, km/s	7.0 to 8.0
Critical diameter of Detonation, mm	2 to 3

Explosive composition		**Lining composition**	
Ballistite base	90 %	Powder metall, %	85 to 90
Sensitizers	10 %	Ballistite base, %	10 to 15
		Density, g/cm	4 to 6

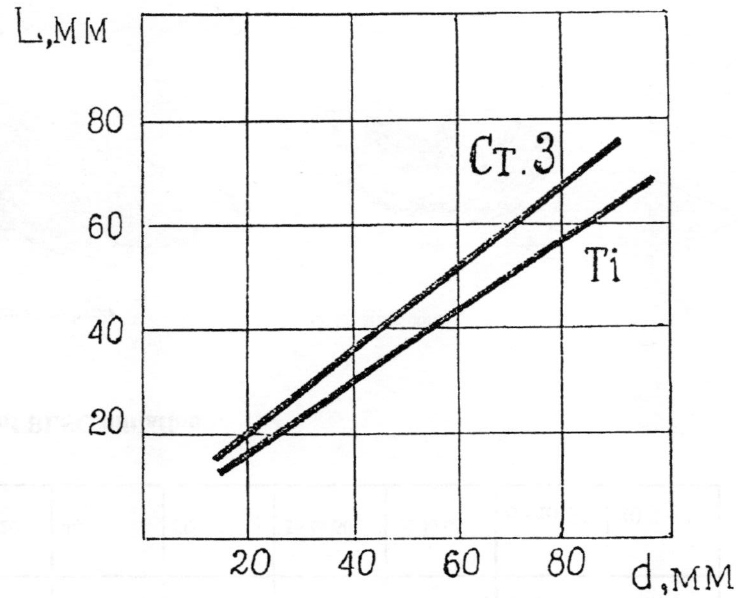

Fig. 5

Depth of L cut vs diameterd of a charge and material nature.

Safety rules when operating material

Additional requirements to shaped charges compared to conventional industrial explosives are absent. Their electrostatic sensitivity is „very low“, i. e. of category 5.

Dropping of BLSC in wood container and without it on to a concrete base to the height of 12m does not cause ignition.

Table 4

Sizes of BLSC according to TU 3-2685-92

BLCS	10	15	20	30	40	50	60	70	80	90	100
Explosive mass without lining, g/m	94	190	360	800	1160	1800	2600	3580	4670	5900	7300
Mass with lining g/m	110	270	500	1120	1700	2700	3800	5250	6850	8700	10700
Thickness of cutted obstacle (steel-20) mm	6 to7	9 to 11	13 to 15	20 to 22	27 to 29	40	50	55 to 60	60 to 65	65 to 65	80

Fig. 6 Methods of BLSC initiation

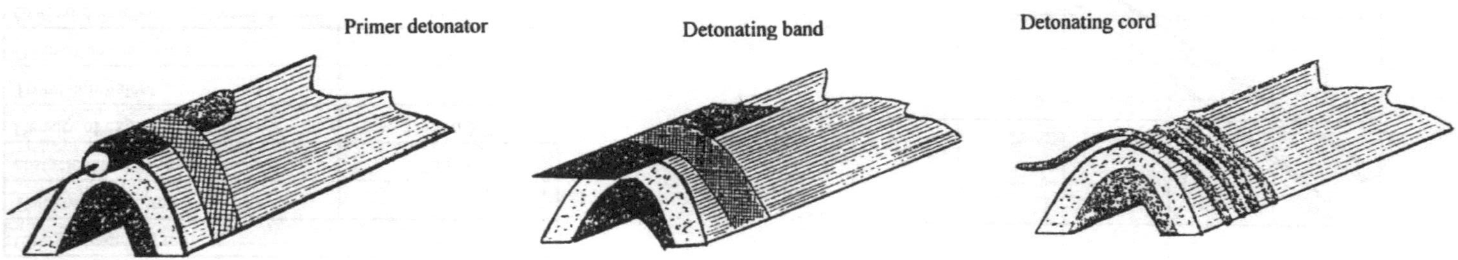

Table 5

Comparative characteristics of linear charges

Charge shape	BLSC	Extended	Shaped cord	Shaped charge
Charge diameter, mm	10 to 100	10 to 25	8 to 30	20 to 30
Thickness of cutted obstacle, mm	7 to 80	8 to 24	6 to 21	20 to 30
Detonation rate, km/s	7.5 to 8.1	8.1	7.9	7.5
Linear mass, kg/m	0.07 to 10.7	--	--	--
Density of explosive g/cm	1.66	1.65	1.6	1.6
Explosive composition	Ballistite	RDX 100 %	RDX	RDX 77 %
Technology	Continuous extrusion of explosive	Batch drawing	Batch extrusion with matalpolymer adhesion	Batch extrusion with metalpolymer adhesion
Cost at equal efficiency	1	1.5 - 2	1.5 - 2.5	

Fig. 7

Technology of BLSC manufacture

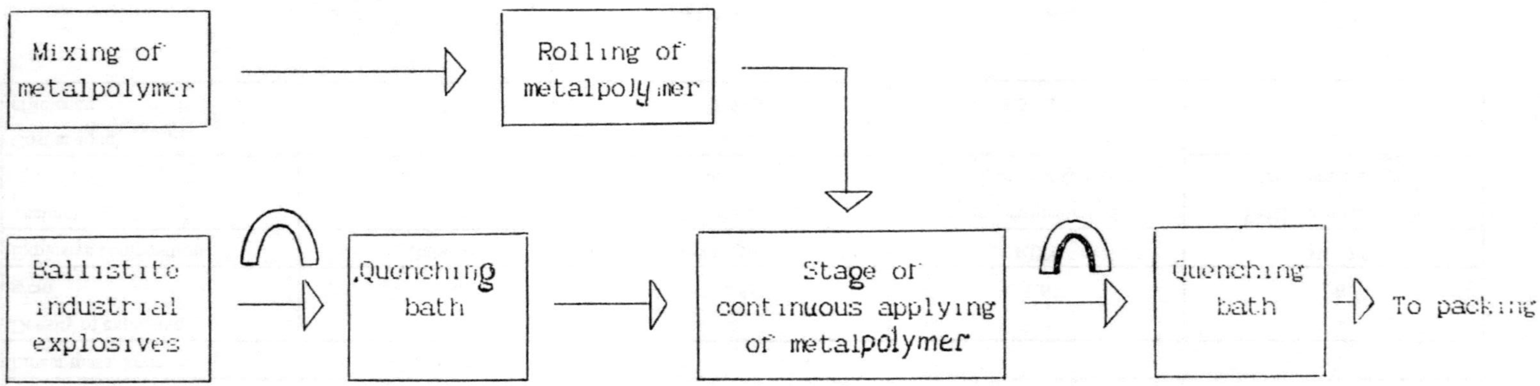

Advantages of BLSC

High efficient continuous technology of BLSC manufacture, low cost compared to similar ones.

High safety in operation, absence of explosion fragments.

Low labour consumption in cutting compared to gas cutting.

Conversion of defence industry.

DEVELOPMENT AND MANUFACTURING OF CHARGES, INCLUDING CHARGES ON BASE OF EXPLOSIVES EXTRACTED DURING UNLOADING OF AMMUNITION, FOR CRUSHING MILL HEAVY SCRAP AND OVERTYRES

S. P. Smirnov, J. S. Kulakevich, E. V. Kolganov, V. I. Abdrahmanov,
A. G. Gorohovtsev, Y. P. Jakovlev

Abstract

The results of the efforts to develop the charges based on the explosives which are recovered from the unloaded ammunition are described in the paper. It is shown by experiment that the recovered explosives, primarily explosive mixtures like TNT/RDX and similar explosives, are advantageous as main charge of the shaped charge which are intended for blasting heavy metal scrap, the metallurgical industrial waste, into fragments. Line case charges based on the recovered explosives and having bulk weight up to 90 kg are developed and intended to blast worn out large-sized tyres into pieces less than 50 kg by weight.

A rational use of the mixed explosives which are recovered from the unloaded ammunition causes us to search for a possible field of application. Naturally, the mixed explosives shall be used in the charges and in the articles which are designed to challenge the problem, ordinary commercial explosives are unlikely to solve. Moreover, the cost of the charges and articles manufactured with the recovered explosives shall be substantially less than the cost of those made with standard explosives, taking into account the expenses to acquire a permit to use the mixed explosives regularly. The current production base where the mixed explosives are prepared and the articles are loaded shall be operated with negligible modifications and no extra investment.

The recovered explosives differ in composition and this difference influences the requirement for efficiency of the charges and the articles in a rather wide range of the filler's energetic characteristics.

H. Krause (ed.),
Conversion Concepts for Commercial Applications and Disposal Technologies of Energetic Systems, 183–190.

We have analyzed the possible fields of application for the recovered mixed explosives while taking into consideration the above requirement, and determined the blast technologies where the charges and the articles are most advantageous. They include the shaped charges for blast fragmentation of mill heavy scraps (up to 150 tons) and the line cased charges for breaking overtyres used in quarry motor-vehicle transport into fragments. The shaped charges are advantageous in blasting fragments of heavy metal scraps due to into metallurgical production. Moreover, currently the mill heavy scraps are accumulated in large amounts and they can't be processed or transported to the blast area because of the large dimensions of a single metal scrap.

According to estimations of experts, the total of mill heavy scraps in Russia amounts to 15-20 million tons, and in the countries of the Common Wealth of Sovereign States it amounts to 45-50 million tons in total. Only about 5 % wt. of the total mill heavy scraps are worked (1).

Till now the use of shaped charges in blast fragmentation was limited because of the lack of shaped charges design approved for regular use which is capable to make holes of the required size and to blast mill heavy scraps into fragments.

The shaped charges applications are restricted to blast oversize pieces into fragments in mineral industry, cutting the thin-walled metal surfaces such as tanks and pipelines, and oil industry (2,3). The weight of the mixed explosives in the shaped charges doesn't exceed 4 kg and the explosive energy is not enough to break a piece of cast iron more than 200-300 kg by weight.

It was necessary to develop the shaped charge which meets the following requirements:

1. The depth of the hole made in heavy steel scrap is from 500 to 1000 mm and the diameter of the hole in a bottom is at least from 30 to 40 mm.
2. Specific output of metal scraps of overall dimensions is at least from 6 to 8 tons, if a single shaped charge is used to break cast iron scraps.
3. No fragments of the shaped charge case are produced in blasting operations.

4. Safe and reliable complies with oecological requirements.
5. The results are reproducible in a wide range of explosives used and with a simplified manufacturing process and available materials used.

The requirements to the shaped charge efficiency were specified according to current practice of breaking heavy metal scrapss into fragments by burning a hole through the scrap with oxygen. Due to fact that the blasting of the heavy metal scrap has to be performed on the storage site, the requirements to the shaped charge safe handling are rigid.

The shaped charge should make uniform holes repeatedly while loaded with explosives off various energetic characteristics, and the recovered explosives used as the main charge should not cause substantial modifications in design of the shaped charge and in its manufacturing and loading process. Moreover, it should be taken into account that the properties of each heavy metal scrap are uncertain depending on different inclusions, type of material and so on.

The efficiency of several shaped charge designs depending on the explosive filler was calculated by a one-dimensional code similar to those in (4-6).

The analyses of calculated data and mock-up tests results substantiated the main design parameters of the shaped charges which differ in efficiency depending on the mass and the type of the explosives filler.

The shaped charge with the hemispheric liner was chosen as a base design because of high reproducibility of the results, when subjected to various disturbance factors such as the constructions components or the main charge defects.

The mock-up shaped charges, 90 mm in diameter and with liner of structural steel 70 mm in diameter, were subjected to the test which uses a steel target with hardness of

300-330 kg/mm^2. The results showed that the calculated depth value agreed well with the experimental data. Thus, for the shaped charge with the liner 2 mm thick and filled with the explosives mixture of density 1.58 - 1.65 g/cm^3 and detonation velocity 7800 - 8000 m/s, the depth of cylinder-type crater 25 - 30 mm in diameter was 160 - 170 mm in comparison with the calculated values of 24 mm and 180 mm correspondingly. An alluminized explosive with the density of 1.7 g/cm^3 and detonation velocity 7000 m/s, which was used as a filler of the charge, made a crater of 36 mm in diameter and of 150 mm depth.
In further research the effects of the explosive energy parameters as well as structural and engineering characteristics of the shaped charges with the explosive mass up to 27 kg were studied.

It was shown that the mixed explosives based on TNT/RDX (TNT 25 - 40 % wt.) produce holes of 500 - 800 mm in depth and 90 - 150 mm in diameter in steel scraps depending of steel type and the weight of the mixed explosive. If the mixed explosives based on TNT/HMX are used, the efficiency of shaped charge increases by less than 10 % where as the cost of the shaped charge increases 3 - 5 times.

The test results for the shaped charges with steel liner 6 mm thick and filled with mixed explosive TNT/RDX, the main charge mass being 25 kg, are given in table 1.

Table 1

N	Composition TNT content %	Density g/cm^3	Metal scrap	Hole making efficiency	
				Depth, mm	Diameter, mm
1	30	1.58 - 1.6	steel	550	120
2	30	1.58 - 1.6	cast iron	500	160
3	24	1.64 - 1.65	steel	600	100
4	24	1.64 - 1.65	cast iron	550	140
5	30	1.6 - 1.62	steel	800	100

Further researches were done to optimize the shaped charge design in order to minimize the mass of the explosives to 18 - 20 kg, to get an approval for regular use and to develop the manufacturing and filling process in a pilot shop in a production plant using available equipment.

The optimized charge with an explosive weight of less than 20 kg (SC-20) was a truncated cone 340 mm in diameter in the bottom and 180 mm in diameter in the top. The length of the charge was 340 mm; the composition was TNT/RDX (40/60), with a density of 1.58 - 1.65 g/cm^3. The auxliary booster (electric detonator) made of desensitized RDX with the density of 1.65 g/cm^3, 80 g by weight and having a channel for detonating primer on the axis was placed in the top of the main charge. Steel liner 6-8 mm thick was a hemisphere 230 mm in diameter at the bottom. The shaped charge was blasted at a distance of 500 - 700 mm from the surface of the metal scrap.

The test results for the shaped charge SC-20 acceptance performed in cinder dump at the metallurgical are given in table 2.

Table 2

N	Number of charges tested	Test site	Efficiency of shaped charges Holes in steel scrap			Slag-cast iron scrap Specific output in tons/ /charge *
			Depth, mm	Diam. in mouth	Diam. at the end	
1	25	Dneprodzerzhinsk	-	-	-	8 - 10
2	10	Beloretsk	-	-	-	10 - 12
3	10	Seversky	500 - 550	110 - 120	40 - 50	-
4	10	Serov **	500 - 550	90 - 110	30 - 40	12 - 15
5	50	Krivoy Rog	550 - 650	120 - 140	40 - 50	10 - 12

* - slag and cast iron heavy scrap

** - alloy steel

It is seen from table 1 and 2 that the shaped charge efficiency is mainly determined by the properties of heavy metal scraps and to a lesser extend it depends of energy of the explosive fillers studied. So the efficiency of the shaped charge SC-20 and its analogs in the detonation velocity range from 7500 to 8000 m/s and the density of 1.6 - 1.7 g/cm^3 will be sufficient. In this connection it is advantageous to use recovered mixed explosives as the main charge filler, if the mixed explosives meet the energy requirements.

As the shaped charge SC-20 which was selected for the base design uses explosive mixture TNT/RDX (40/60) and the share of such explosive mixtures in the total amount of the recovered ammunition is relatively small, the target mixed explosive can be produced by thickening less concentrated cyclotols.

It is reasonable to use desensibilized mixed explosives similar to „Comp. B“ as a filler and to use formulations which contain aluminum and are similar to compositions such as „HBX-1“, HBX-3“, „HZ-1“, because they are widely used and offer reasonable energetic characteristics as evidenced by results of mock-up tests.

A second economic and oecological problem to be resolved in Russia and in the countries of the Commonwealth of Sovereign States relates to recycling of worn out large tyres and overtyre-rubbers which contain 50 % and 95 % of natural rubber correspondingly.

Dumps of worn out tyres in quarries, minerals separation plants take up large areas and deteriorate oecology. According to expert estimations, a total mass of large tyres and overtyres up to 65 thousand tons per year goes out of use in Russia.

As far as we know, there is no reliable and cost effective method for mechanical processing of worn out large tyres and overtyres in the world. Preliminary analysis has shown that blast fragmentation of tyres is cost effective, if a rational engineering approach is used.

Experimental testing of alternative methods to blast worn out tyres, type of the charge, layout of the charges, type of the explosive was carried both for medium-sized and full-scale large-sized tyres. The main part of research was carried for overtyres (3.5 ton by weight; 3600 mm in diameter) with textile cord and for similar import tyres with metal cord.

The attempts to cut tyres using linear shaped charges (LSC N°6) failed. The resulting cutting depth was less than 25 - 30 mm. After the paste explosive charges in the form of tapes 20 - 30 mm wide and about 20 mm long have been arranged on the protector and the sides and blasted, only traces of blast wave are found on the surface of the tyre. Further, the possibility of using hydroshock which is produced by explosives blasting was studied. It was shown that the line cased charges up to 90 kg in weight placed inside the tyre in bed of water, that is using hydroshock, are capable to separate rubber from cord. As result, pieces of 50 kg by weight are produced. Emulsion explosive „Poremit" was used. Implementation of this method in quarries is complicated by the difficulties of tyre hermetic sealing, pumping water and so on.

In this connection an alternative scheme was developed which is based on blast effect of the explosive in confined air. Blast energy effect of spaced apart charges and linear toroidal charges placed inside the tyre on the fragmentation behavior was studied. The experiments were carried to particularize the layout of charges inside the tyre, to optimize the size and the weight of the charge and to correct the initiation circuit. The charges of ammonal and ammotol from 35 to 100 kg in weight were used in this experiment.

The experiment showed that cutters of special design are required. The tyres are placed on the cutters and the bead rings with metal cable embedded are cut out. Metal cables account for 75 % wt of bead rings.

Large pieces of rubber were broken completely by superimposed line cased charges. The rubber pieces of the following size are produced: 60 cm long, up to 17 cm thick, up to 25 cm wide. The pieces of rubber are suitable to be worked into crumb.

As the energy characteristics of the explosives have considerably less influence on fragmentation of the large-sized tyres than the charge design and initiation techniques have, it seems reasonable to use mixed explosives prepared on the basis of explosives, which are recovered from unloaded ammunitions.

Conclusion

Base design of the shaped charges SC-20 is developed which complies with the requirements to the articles of consumption. A series of tests were performed in six metallurgical plants in Russia and Ukraine. CS-20 has passed the acceptance test and is recommended for regular use.

It is shown by experiment that the recovered explosives including aluminum-containing explosives are suitable as the main charge for effective blasting fragmentation of steel and cast iron scraps.

Line cased charges filled with commercial explosives up to 90 kg by weight, density 900-1300 kg/cu.m, detonation velocity 4000 - 5000 m/s are developed and used for blasting overtyres into fragments of the predetermined size.

Reference

1. Schubert, H. Preparing of metal secondary raw material (resources, classifications, reductions in size). Translated from German - Moskow: Metallurgy, 1989, 360 pages
2. Bickbaev, A. Z. Blast energy used for tank stock reconstructions. Thesis - Ufa, 1990, 197 pages
3. Blast shooting equipment. Manual. M. Nedra, 1990, 278 pages
4. Fizica Vzryva. Moskow: Nauka, 1975 p. 3-67-440, 704 pages
5. Kinelovsky, S. A. Trishin, Yu. A. Cumulation physical aspects. Fizika Goreniya i Vzryva. - 1980. - v. 16, N 5. - P. 26 - 40
6. Tomashevich, I. I. Target penetration by high velocity flow of elongated rods. Fizika Goreniya i Vzryva. - 1987. - v. 23, N. 2 - P. 97 - 101.

CHEMICAL AND BIOLOGICAL PROCESSES FOR DECOMPOSITION OF NITROCELLULOSE-BASED PROPELLANTS

André DAHN
BC Berlin-Consult GmbH
Am Karlsbad 11
W-10785 Berlin
Germany

1. Introduction

A process for chemo-biological disposal of propellants coming up from dismanteling of ammunition was examined as to its suitability for large-scale technological use within the scope of a study funded by the BWB Koblenz and was assessed both technologically and economically.

Proof of the functionability of this method of waste disposal has already been furnished for a number of propellants (see Table 1).

The process involves treatment of aqueous suspensions of propellants in the following steps as shown in figure 1:

- hydrolysis and denitration of propellant constituents;
- biological conversion of hydrolysate substances;
- post-treatment of process water.

H. Krause (ed.),
Conversion Concepts for Commercial Applications and Disposal Technologies of Energetic Systems, 191–212.

Further studies were necessary to show to what extent an aerobic biological stage can improve the quality of the process waste water with respect to nitro and amino aromates. For examining the biological degradation and the utilization of the residual substances, pollutant tests in accordance with Water Purification Regulations (AbfKlärV) were carried out on the excess sludges from the biological stage.

In addition, a thermal hydrolysate treatment stage was linked between the hydrolysis stage and the biological treatment stage with the goal of achieving additional improvement in the COD quality of the effluent so that several process steps in the post-treatment of the process water could be avoided.

2.2. PRELIMINARY REMARKS ON FURTHER INVESTIGATIONS

For the experimental goals mentioned in Section 1, continuous experiments were carried out both in the bench scale and in pilot plant. In contrast to earlier experiments, production of hydrolysates took place exclusively in the hydrolysis stage of the pilot plant. For the tests for nitro and amino aromates, NDT3 hydrolysate was employed in the laboratory plant and Wufl hydrolysate in the pilot plant. The stabilizers diphenylamine (DPA) and Zentralit I and II were difficult to degrade and used as monitoring substances for nitro and amino aromates.

For investigations on the further reduction of the COD, hydrolysates of the Wufl propellant were treated at different temperatures and thereafter subjected to biological post-treatment. The specimens for the nitro and amino aromate tests were taken in the laboratory and pilot plants one week after setting of a new operating point.

The results of investigations obtained from the laboratory and pilot plants are presented in the following.

2.3. RESULTS

2.3.1. Decomposition of nitro and amino aromates in the hydrolysis stage of the pilot plant

Table 4 shows the degradation rates for propellant substances after hydrolysis of the propellants NDT3 and Wufl.

The input concentration of the NDT3 substances was already shown in Table 2 and is taken over here in an abbreviated form.

After hydrolysis, the nitro aromates in the NDT3 propellant are eliminated almost completely. Only 0.4 mg/l of the reaction product 2-nitrotoluene was found in the hydrolysate. The propellant constituents DNT and TNT are no longer present. The stabilizers Zentralit and diphenylamine are in each case reduced to a residual percentage of 0.4%.

The residual percentages are of considerable relevance with respect to the further consideration of water-polluting substances. The other measured substances such as NAT, 1-phenyl-benzotriazole and N-phenylaminobenzene are no longer of any significance as will be seen in the following Section.

2.3.2. *Biodegradation of hydrolysis reaction products*

Table 5 shows the degradation results obtained in the laboratory plant.

The substances nitrotoluene and nitroaminotoluene still present in the hydrolysate are degraded to below detectability.

Depending on the propellant dosage, DPA from Wufl hydrolysate as well as Zentralit 1 and 2 from NDT3 are detectable up to the anoxic stage of the biological line.

Since the first substances used by the denitrificating micro-organisms are the easy-to-degrade substrates, there is not enough oxygen present in the anoxic stage for hard-to-degrade substances such as DPA and Zentralit. Degradation of the stabilizers is thus not complete in the anoxic stage. Further decomposition starts to take place in the aerobic stage. In this connection, DPA was reduced to below the detectable level.

Decomposition of Zentralit 1 and 2 amounts to about 70%.

Table 6 shows the degradation degrees achieved in the anoxic stage and in the whole biological process of the pilot plant. In order to clearly show the nitrite and nitrate limitation in the anoxic stage, these substances are also included in the table. Nitrites were converted to N2 down to detectability.

The nitrogen concentration in the effluent of the anoxic stage likewise is inside the range of accuracy of the measuring process. In as far as there is no urea added in the process, the ammonium concentrations are very low. At such low effluent concentrations the ammonium degradation rate is not critical as regards water conservation.

The utilization of nitrites and nitrates was better in the pilot plant than in the bench scale experiments. This can serve as an explanation for the fact that hard-to-degrade substances such as N-phenyl-1,2-diaminobenzene and 1-phenyl-benzotriazole as well as diphenylamine could be broken down to below the detectable limit. For the stabilizers, Zentralit 1 and 2, too, a higher degradation rate arises in the anoxic biological stage at 80%. The degradation can be increased by an aerobic biology stage to 83% for Zentralit 1 and 98.7% for Zentralit 2. The effluent concentrations at 90 μg/l and 8 μg/l respectively thus are at an acceptable level as regarding water conservation regulations. With respect of DPA, it is even possible to get below the statutory effluent discharge limit of 20 μg/l in water protection zones after completion of the biological stage.

The subsequent stage for process-water treatment serves primarily for reducing the COD. In order to avoid this post-treatment stage, the COD of the effluent must be lowered still further by process optimization after the hydrolysis stage. This is treated in more detail in Section 2.3.5.

2.3.3. Heavy metal concentrations in the process waste water of the pilot plant

When using the Wufl propellant, no direct pollution of the process water by heavy metals was to be expected. Nevertheless, control measurements of the biological effluent had to be taken in order to establish to what extent processes of re-dissolution from plant parts (in particular the fresh water input line) impair the quality of the water. The analysis results show that there is no particular worsening of the biological effluent as a result of heavy metals. The effluent concentrations are listed in Table 7.

The extent to which heavy metals are present in the bio-sludge is shown in the following Section.

2.3.4. Results of sludge tests

With respect to the subsequent possible utilization of the bio-sludges coming up in the abatement plant as a fertilizer substitute, surplus sludge tests were carried out for all pollutants in accordance with the Waste Purification Regulations (AbfKlärV) and also for assessment of the parameters relevant for the sludge quality. The results can be seen in Tables 8 and 9.

From the results the following can be stated:

1. All limits in accordance with the Waste Purification Regulations (AbfKlärV) are more than adhered to. This applies for all heavy metals and all PCB isomers.
2. The content of Zentralit 1 and 2 is low. The elimination of the stabilizers in the biological stage prooves biodegradation and not adsorption on the sludge. Diphenylamine and aminobenzenes are no longer detectable.
3. Residual organic acids are present in the sludge.

Post-treatment of the sludge by aeration (aerobic stabilization) may improve the quality of the sludge. The rotting ability of the sludge is then lost. The content of organic acids and Zentralit 1 and 2 is further lowered. The organic fraction in the sludge can be reduced to 44%. The heavy metal and PCB contents are only increased insignificantly due to the loss of mass. After sludge treatment, they do not exceed the limits in accordance with the Waste Purification Regulations (AbfKlärV).

The sludge can presumably be used for soil recultivation at shut-down open pit mining locations.

2.3.5 Results obtained from investigations aimed at optimizing COD reduction

25% of the propellant mass consists of C-H compounds. In single-base propellants, these compounds consist almost exclusively of nitrocellulose which do not completely lose their high-molecular character after hydrolysis. A large percentage of the organic compounds still have a molecular weight of > 1,000 g/mol after hydrolysis. Degradation of these compounds and thus corresponding COD reduction only takes place very slowly. Biological degradation of these substances can be accelerated by thermal treatment which from the technological point of view of the process takes place as a hydrolysate post-treatment step.

Fig. 2 shows the biodegradability as a function of the thermal pre-treatment at a temperature ranging from 120 to 180 ° C, taking a propellant hydrolysate as an example. A clear improvement in the degree of biodegradation will become evident at reaction temperatures above 150 ° C.

Application of hydrolysate post-treatment has the following advantages in the overall conversion process:

1. The improvement in biodegradability permits saving of co-substrates in the denitrification stage of biological treatment.
2. COD, nitrite and nitrate contents are lowered. The predilution of the hydrolysates is decreased markedly.
3. The percentage of circuit water used for hydrolysate pre-dilution can be increased. The need of fresh water is decreasing.
4. It will probably be possible to use more cost effective processes for the process water post-treatment stage.

Fig. 3 shows the mass balance of the propellant disposal process with hydrolysate post-treatment (T = 180 ° C). All in all, an overall degradation degree of 97% is achieved for the propellant.

3. Results of further investigations

3.1. HYDROLYTIC TREATMENT OF RDX

Jones [1] and Hoffsommer [2] had already carried out investigations on the alkaline decomposition of RDX. NO2-, N2, NH3, N2O, HCOOH, CH2O and H2 were detected as reaction products.

3.2. HYDROLYTIC TREATMENT OF HMX

Just like RDX, HMX can be transformed largely by alkaline hydrolysis. Results of investigations were provided by Croce [3]. Kinetic investigations on alkaline hydrolysis of RDX, HMX and TNT are presently being carried out by Heilmann [4].

3.3. TREATMENT OF TNT

In the course of the treatment of TNT various anionic transformation products are formed at room temperature and pH 7. However, no certain information on a further breakdown is available. The transformation products observed bear a negative charge. Bernasconi [5] and Urbanski [6+7] indicate that all these reactions are completely reversible. The rates of reaction are in the range of those of normal acid/base reactions.

However, there does not exist any data on the extent to which the products of the reactions react further. To this end further orientating experiments were carried out with TNT dissolved in water by Hesselmann [8] at the behest of BC. The following can be stated as a result of the hydrolysis experiments:

1. Not only do the reversible transformation reactions as described by Bernasconi take place but also irreversible decomposition and/or transformation reactions. Thus the reactions are more complex than originally assumed.
2. A large number of these reactions generate polar substances which in polar eluants (water/acetonitriles) elute before TNT from the non-polar HPLC acid. It can therefore be assumed that these products bear a negative charge and/or have a lower molecular mass than TNT.
3. Since no fine turbidity or precipitation was observed in the hydrolysed solutions, the metabolites thus produced are at least as soluble in water as TNT. Due to the elution characteristic (see b), the hydrolysis products should in general be more soluble in water than TNT.

The subsequent photolysis of TNT and its hydrolysis products showed that decomposition of TNT as well as its reversible and irreversible hydrolysis products is possible to a wide extent on exposure to UV-light. This can be concluded from the almost complete disappearance of UV-absorbent substances.

In addition, it turned out that, in the investigation of double-base propellants, the tested propellant NDT3 also contained TNT in addition to the familiar explosive components. Control analysis after hydrolysis proved that the TNT fraction in the propellant was completely eliminated. In retrospect, it can be concluded from this that pure hydrolysis of TNT is also possible. In the course of further investigations, it would make sense to look in more detail into the safety-relevant aspects of hydrolysis of explosives and thereafter to examine the biodegradability of the hydrolysates. The conclusions drawn from this would then serve as a reference for the disposal problems touched upon in Section 6.

4. Prospect for other type of waste disposal

Left-overs of explosives such as TNT and Hexogen were dumped and tipped at the time of the Third Reich at the locations of what used to be armament plants. The result of this method of "waste disposal" is wide-spread soil contamination as well as inestimable contamination of the ground water and percolating water. Some locations are thus urgently in need of decontamination. These include the site at Elsnig/Torgau, in addition to sites in Hesse (Stadtallendorf and Hessisch-Lichtenau). At the first-mentioned location, TNT and Hexogen are stored in a dump right on the site of a water protection area. The TNT dump is washed directly by the aquifer. The eluates arising contaminate the ground water which is required for drinking water purposes for the Halle/Leipzig area. Hitherto this contaminated water was purified by activated charcoal filters. In the medium term, it can be expected that the TNT-contaminated soil will likewise be cleaned by washing. The wash water arising in the course of this soil wash process can without doubt likewise be purified with activated charcoal filters.

In addition, in the course of soil washing, highly contaminated fine sand fractions and explosive-water suspensions arise. These residues (contaminated fine sand fractions, TNT-water mixtures) could be specially treated in a chemo-biological plant, i.e. decontaminated by this means. In this connection, it is important to establish the extent to which charged activated charcoal could be regenerated with the aid of chemo-biological methods and re-used for purification of the ground water. The corresponding fields of use and types of processes intended for this purpose are shown in Fig. 4.

5. Summary

The chemo-biological process is conceived for the treatment of organic compounds which are hard to access by biological means alone. These include explosive solids which by this method can be specifically decomposed in an aqueous medium without hazard. Results arrived at hitherto permit the conclusion to be drawn that other nitro-compounds too (e.g. TNT or Hexogen) could likewise be successfully disposed of. The field of use would in particular also include treatment of left-overs from soil decontamination in the field of cleaning up of historical military contamination.

To investigate this new field of use, investigations on safety technology as well as hydrolysis and biodegradation investigations must be carried out on a semi-technical scale.

For experimental runs, a pilot plant is available for this purpose. The goals of investigations should include:

- treatment of TNT and Hexogen;
- treatment of TNT and Hexogen-contaminated concentrates (silt) from the abatement of historical contamination;
- treatment of explosives-contaminated soil.

Bibliography

[1] Jones, W.H. (1953): Mechanism of the homogeneous alkaline decomposition of cyclotrimethylene-trinitramine. Kinetics of consecutive second and first-order reactions. A polarographic analysis for cyclotrimethylene-trinitramine. J. Am. Chem. Soc. 76: 829 - 835.

[2] Hoffsommer, J.C., Kubose, D.A. and Glover, D.J. (1976): Kinetic isotope effects and intermediate formation for the aqueous alkaline homogeneous hydrolysis of 1,3,5-triaza-1,3,5-trinitro-cyclohexane (RDX). J. Phys. Chem. 81: (5), 380 - 385.

[3] Croce, M. and Okamoto, Y. (1979): Cationic mecellar catalysis of the aqueous alkaline hydrolysis of 1,3,5-triaza-1,3,5-trinitrocyclohexane and 1,3,5,7-tetraaza-1,3,5,7-tetranitrocyclooctane. J. Org. Chem. 44: (13), 2100 - 2103.

[4] Heilmann, H.M. (1993), UCLA - Civil Engineering Department, 4145 Engineering, private communication.

[5] Bernasconi, C.F. (1971); Kinetic and spectral study of some reactions of 2,4,6-trinitrotoluene in basic solution. I. Deprotonation and Jankovsky complex formation. J. Org. Chem., 36: (12), 1671 - 1679.

[6] Urbanski, T. (1984): Chemistry and technology of explosives. Vol. 4, Pergamon Press, Oxford.

[7] Urbanski, T. (1967): Chemistry and technology of explosives. Vol. 4, Pergamon Press, Oxford.

[8] Hesselmann, R. (1993), Experimental Report "Alkalische Hydrolyse und Photolyse von TNT", EAWG Eidgenössische Technische Hochschule Dübendorf, Switzerland.

Table 1

Name type of propellants Components	TM I solid % by mass	TM II solid % by mass	TM III NDT-3 % by mass	TM VI Pyroxilin-Powder % by mass	TM VII DG 3/1 % by mass	TM VIII Sa 2P 15 % by mass	RS I-12 K % by mass	Wufl Powder % by mass	NGSP Powder % by mass
Nitrocellulose	57,5	57	56,3	96	60	60	56,3	97	81
Nitroglycerine	26,7	28	26,6	-	35	28	26,3	-	14
Dinitrotoluene	8,5	11	6,4	-	-	5,6	11,4	-	2
Zentralit	2,9	3	3,4	-	4	1,7	2,9	-	2
Diphenylamine	-	-	-	1,5	-	-	-	2	-
Graphite	-	-	-	-	-	-	-	1	2
Lead	1,4	-	-	-	-	1	-	-	-
Titanium	-	-	-	-	-	1,5	-	-	-
Water	0,5	-	-	-	-	-	-	-	-
Dibutyl phthalate	1,5	-	4,9	-	-	-	-	-	-
Vaseline	1	1	-	2,5	1	-	1,2	-	-
Calcium carbonate	-	-	-	-	-	-	0,9	-	-
Smoothing paste	-	-	-	-	-	-	1	-	-
Trinitrotoluene	-	-	2,4	-	-	2,2	-	-	-
Main components	95,6	99,0	92,7	96,0	99,0	95,3	96,9	97,0	99,0
Resid. fractions	4,4	1,0	7,3	4,0	1,0	4,7	3,1	3,0	1,0
Total	100	100	100	100	100	100	100	100	100

Table 2

Mass balance of chemo-biological propellant disposal process

Substance in propellant NDT-3	Initial solution (mg/l)	After hydrosysis as undiluted hydrolysate, laboratory plant (mg/l)	After anoxic biological stage converted as concentrate (mg/l)
2,4 DNT	4.640	8	)
2,6 DNT	1.150	3,4	)
3,4 DNT	18	< d. t.	) d. t.
3,5 DNT	5	< d. t.	)
Σ DNT	5.813	11,4	)
TNT	574	< d. t.	
NT	0	26,5	1,8
NAT, ANT	0	20,0	2,0
NMT	0	4,0	< d. t.
Zentralit	3.322	36,6	11
Anilines	0	4,76	1,1

d. t. (detectibility threshold) = 3 μg/l

Table 3

Degradation degrees in anoxic biological stage

ropellant	Components	Influent μg/l	Effluent μg/l	Degree of elimination
T-3	Zentralit	1.660	549	67
	Σ Nitrotoluene	1.707.	82	95
	Σ Anilines	1.107	142	87
	Methylbenzofurazane	116	47	59
	1,2-Benzene-dicarboxylicacid butyl ester	12	n. d.	100
	Tetradecane	15	n. d.	100
	Hexadecane	34	n. d.	100

n. d. = not detected

Table 4

Degradation degrees in hydrolysis process and reaction products

Substande in propellant NDT-3	Initial solution (mg/l)	After hydrolysis stage als undi-luted hydroly-sate, pilot plant (mg/l)	Degradation rate (%)
Σ DNT	5.813	< d. t.	100
TNT	574	< d. t.	100
NT	0	0,4	
NAT, ANT	0	28	
Σ Zentralit	3.322	16,2	99,6
Anilines	0	0	
Wufl			
DPA	2.000	8,6	99,6
Benzotriazol-1-phenyl	0	0,077	
Σ Zentralit	0	6,6	
N-Phenyldiaminobenzene	0	0,086	

d. t. (detectability threshold) = 3 µg/l

Table 5

Degradation results from biological stages of laboratory plant (input material NDT-3)

Components in NDT-3 hydrolysate	Influent into first biological stage µg/l	Effluent of anoxic stage µg/l	Degree of degrada-tion %	Effluent of aerobic stage µg/l	Degree of degrada-tion %
2 Nitrotoluene	83	< d. t.	> 97	< d. t.	< 97
Σ Nitro-amino-toluene	1.120	< d. t.	100	< d. t.	100
Zentralit	1.717	1.131	34	523	70

d. t. (detectability threshold) = 3 µg/l

Table 6

Degradation results from biological stage of pilot plant

Components in Wufl hydrolysate	Influent to biological treatment mg/l	Effluent of anoxic stage mg/l	Degree of degrada-tion %	effluent from aero-bic stage mg/l	Total degree of de-gradation %
Ammonium/ Ammonia	0,6	0,2	66	0,2	66
Nitrate	265	0,13	99,99	42,2	84
Nitrite	1.097	< 0,05	100	< 0,05	100
Diphenylamine	2	< d. t.	100	< d. t.	100
N-methyl-diphe-nylamine	0,006	0,006	0	< d. t.	> 50
N-phenyl-1,2-di-aminobenzene	0,006	< d. t.	> 50	< d. t.	> 50
Zentralit 1	0,53	0,15	72	0,09	83
Zentralit 2	0,59	0,012	80	0,008	98,7
1-Phenylbenzo-triazole	0,006	d. t.	> 50	< d. t.	> 50

d. t. (detectability threshold) = 3 µg/l

Table 7

Heavy metal contents in effluent from biological stage

Element	Effluent content (µg/l)
As	< 2,5
Cd	< 10
Co	< 10
Cr	14,7
Cu	47,9
Hg	< 1,25
Ni	< 20
Pb	< 100
Sn	< 100
Zn	49,1

Table 8

Results of sludge analysis in accordance with Waste Purification Regulations

Substance	Dimensions	Waste Purification Regulations	Analysis of excess sludge
Parameter pH value	--	< 6	7,7
Dry matter (D. M.)	%	N. A.	35 plus dehydration
Ignition loss	% D. M.		66
Nutrients			
Total N	% D. M.	N. A.	2,190
NH4-N	% D. M.	N. A.	0,077
Phosphate	% D. M.	N. A.	8,090
Potassium oxide	% D. M.	N. A.	1,220
CaO	% D. M.	N. A.	4,880
MgO	% D. M.	N. A.	0,350
Heavy metals			
Lead	mg/kg - D. M.	900	58,2
Cadmium	mg/kg - D. M.	10	0,45
Chromium	mg/kg - D. M.	900	29,3
Kupfer	mg/kg - D. M.	800	150
Nickel	mg/kg - D. M.	200	7,33
Mercury	mg/kg - D. M.	8	1,57
Zinc	mg/kg - D. M.	2.500	1.120
AOX	mg/kg - D. M.	500	70
PCB			
28	mg/kg - D. M.	0,2	0,101
52	mg/kg - D. M.	0,2	0,047
101	mg/kg - D. M.	0,2	0,102
138	mg/kg - D. M.	0,2	0,151
153	mg/kg - D. M.	0,2	0,156
180	mg/kg - D. M.	0,2	0,091
Total	mg/kg - D. M.	N. A.	0,648

Table 9

Further results from sludge analysis

Substance	Concentration (mg/kg)
DPA	< d. t.
N-Methyl-DPA	< d. t.
Zentralit 2	0,5
N-Phenyl-diaminobenzene	< d. t.
Zentralit 1	8,7
1-Phenyl-benzotriazene	< d. t.
1,2 Phthalic acid alkyl ester	23
Heptadecene	0,25
Total alkane acids	5,7
Hexadecanic acid	0,5
Total paraffins	2,5
N-Octyl ester palmatinic acid	1
Indene	< d. t.
Tridecanone	< d. t.
9-Acridinamine	< d. t.

d. t. (detectability threshold) = 0.25 mg/kg

input material	prop. 333 kg/h H_2O 3,3 m^3/h	NaOH 208 kg/h	Dilution water max. 36 m^3/h	Glucose Ethanol	Ozone, H_2O_2 10 kg/h	
process step	prop. dosing	hydrolysis denitration	hydrolysate post treatment	biological treatment	effluent post treatment wet oxid.	biol. post treatment
residual matter		hydrolysate	hydrolysate waste water	sludge 10 kg/h dry matter waste water	effluent	
waste water		3,3 m^3/h	3,3 m^3/h 10–40 m^3/h	10–40 m^3/h	10–40 m^3/h	
COD (g/m^3)		62000	52000 4000	1000	<200	
Nitrite (g/m^3)		25000	20000 1500	0	0	
Nitrate (g/m^3)		20000	20000 1500	10	10	
BOD_5 (g/m^3)					<15	

Fig. 1 Block diagramm - chemo-biological propellant disposal

BC Berlin

Fig. 2: Biodegradability of Hydrolysate vs. reaction temperature of post treatment

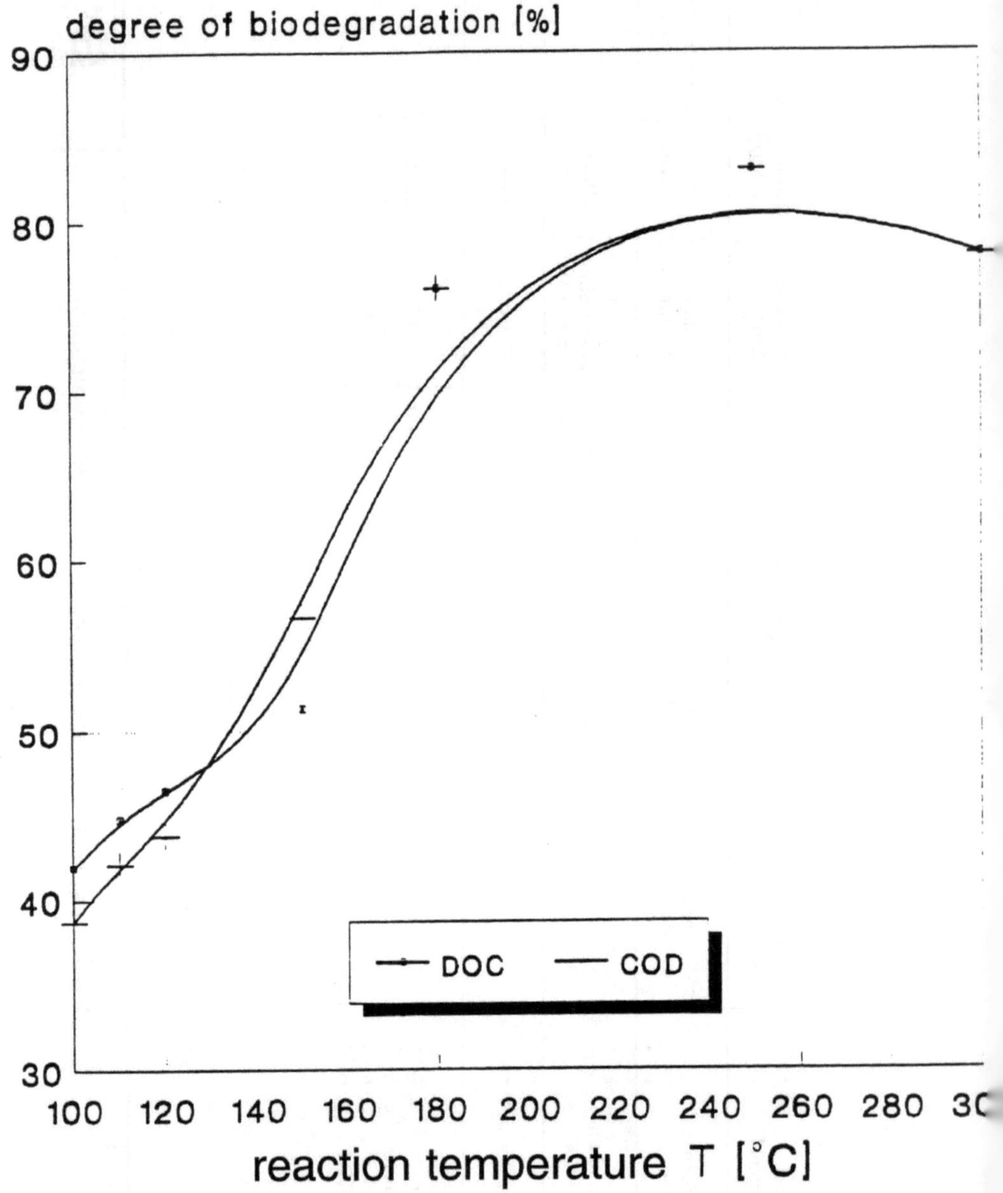

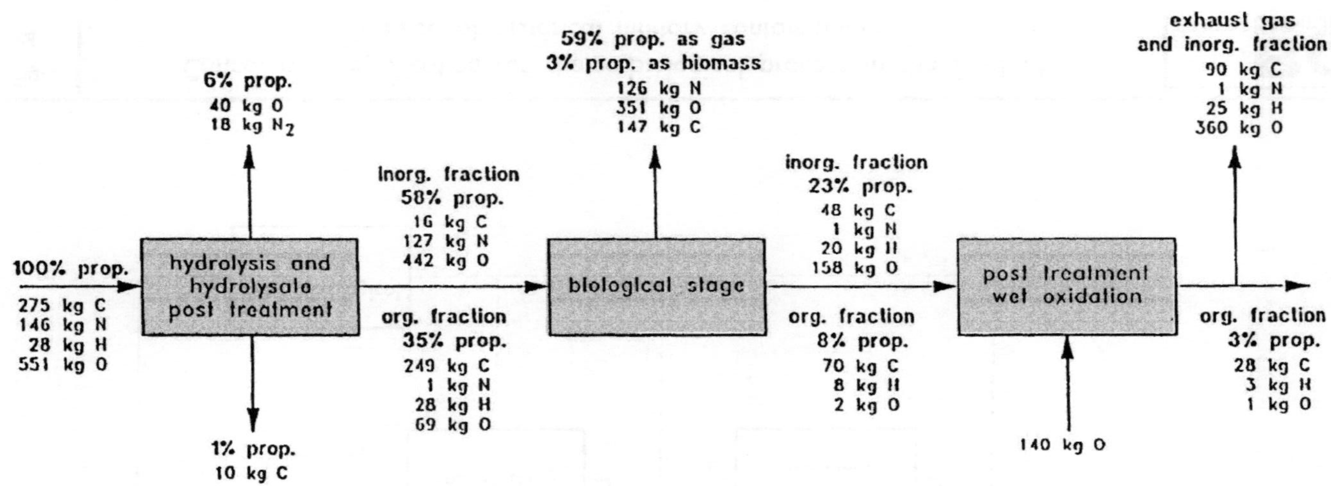

Fig. 3 Predicted mass transfer in the case of a propellant abatement plant with a hydrolysate post treatment stage (T=180°C)

BC Berlin

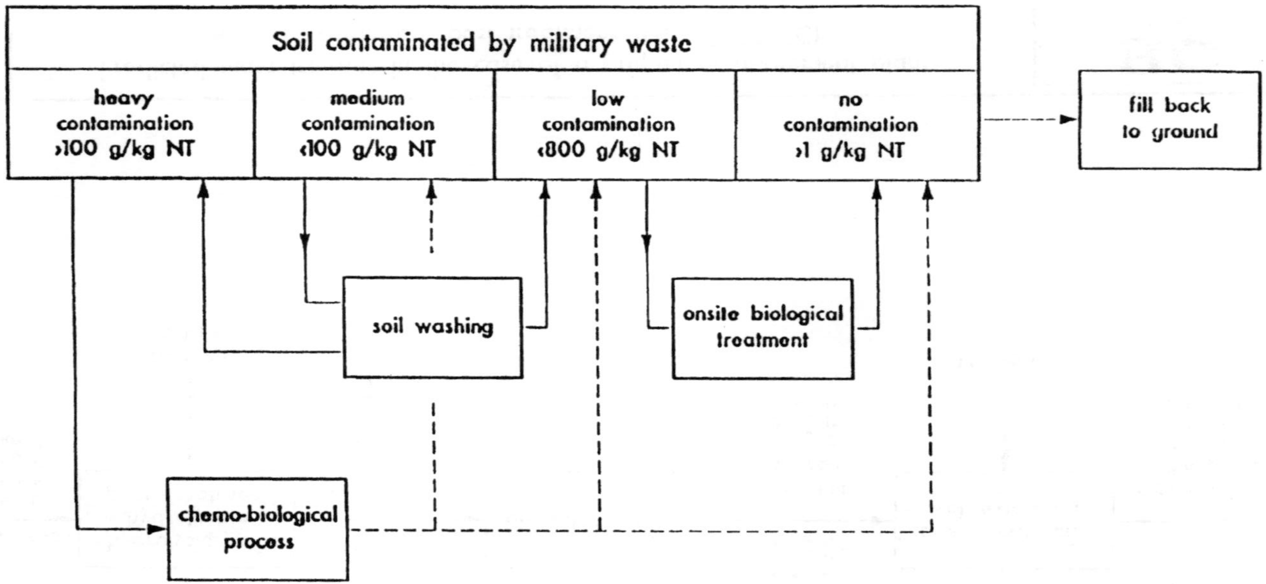

Fig. 4 Conceivable application for chemo-biological process in the field of clean-up of historical military contamination

USE OF POWDERS IN NATIONAL ECONOMY

E. F. Okhrimenko, N. G. Ibragimov, E. Kh. Afiatullov, L. S. Lukin,

Yu. M. Yukov, I. P. Ivanova

The Research Institute of Polymeric Materials (NIIPM) carries out promising investigations aiming to develop and introduce highly effective methods and means for usage of powders explosion energy in various branches of national economy.

In our opinion, the most interesting trend is the development of new industrial explosives and, first of all, industrial explosives designed for work in drowned wells, in aggressive media and under different climatic conditions.

There are complex technical requirements to such types of explosives, and two main features can be marked out:

- insolubility in water and no hygroscopicity,
- low price that makes this explosive competitive with the known analogues in use.

In this respect the usage of ballistite materials of high detonating properties of extreme waterstability and resistance to other media seems to be rather perspective.

At present in our country there is accumulated a huge reserve of ballistite powder and fuel „outdated" either due to their tactical and technical characteristics and being of no interest for their direct use, or due to their shelf life run out. The destruction by direct burning is harmful from the economical point of view and inexpendient from the economical point of view. Therefore, the search for peaceful and useful methods using such powders represents an actual problem.

Starting from this and in accordance with the programs of military industry conversion the NIIPM has developed a new, highly effective waterstable industrial explosive, Dibasit - on the basis of outdated ballistite powders and fuels. The main task of developing of this

H. Krause (ed.),
Conversion Concepts for Commercial Applications and Disposal Technologies of Energetic Systems, 213–219.

explosive was to impart to hard ballistite powders a certain number of service and safety features competitive with those of the corresponding explosives used at the present time.

On the basis of research works there were developed technical and technological documents, and pilot lots of Dibasit have been manufactured and accepted by tests under industrial conditions.

The main explosive and physico-chemical characteristics of the new industrial explosive Dibasit are shown in Table 1 in comparison with the standard explosives.

Table 1

MAIN EXPLOSIVE AND PHYSICO-CHEMICAL CHARACTERISTICS OF EXPLOSIVES

CHARACTERISTICS	Dibasit	Granipor ГП-1, ГП-3	Grammonit 30/70	Granulotol
Waterstability, day, min.	60	30	0,5 *	30
Detonation rate, km/s				
- water free	2.8 - 3.5	3.4 - 3.9	4.0 - 4.7	4.0 - 4.6
- water-filled charge	5.8 - 6.0	5.3 - 6.2	5.5 - 6.0	5.0 - 5.5
Detonation critical diameter, mm				
- open charge free of water	100 - 120	100 - 120	40 - 60	60 - 80
- water-filled charge	< 25	80 - 100	10 - 15 **	25 - 30
Oxygen balance, %	minus 45	minus 50	minus 45,9	minus 74
Explosion heat, kcal/kg	800 - 900	800 - 900	870	870
Gas volume, l/kg	850 - 900	850 - 900	800	800
Capacity, acg. to GOST 4546, cub.cm	280 - 300	270 - 300	330 - 340	285 - 295
Bulk density, g/cub.cm	0.85 - 0.95	0.8 - 0.85	0.85 - 0.9	0.9 - 1.0

* - for limitedly drowned wells with unrunning water,

** - in steel case

The Dibasit is manufactured only on the basis of ballistite compounds with the additives which improve explosive and physico-chemical characteristics increasing safety of works. By the danger degree of treatment and conditions of the use Dibasit belongs to class 1, subclass 1.1, compatibility group A and is designed for carrying out explosions during open mining in the wells of any degree of drowning including the wells with running water, under any climatic conditions. It is manufactured as the grains of arbitrary form, maximum size is up to 20 mm. Dibasit was tested under various mining-geological conditions while exploding hard and medium-hard rocks. Dibasit is admitted for regular use by the Decision of GOSGORTEKHNADZOR of Russia HP 11 - 22/579. In contrast to previously designed explosives State Committee of sanitary epidemiological inspection (Sanitary certificate HP 232 of 17.12.193) has admitted Dibasit for manufacture, delivery and use.

The new industrial explosive Dibasit is manufactured as per process line for manufacturing ballistite powders and fuels and includes the following main steps:

- producing of the mass or grinding of finished powder components;
- rolling of the powder components with sensitizer and producing of Dibasit in the form of „tablets" or granules;
- worm pressing of Dibasit tablets to produce finished granulated product;
- dosage, packaging, marking and completing the lots.

The present scheme of explosive Dibasit manufacture was realised by three enterprises.
The second important direction of the NIIPM activity on peaceful use of powders energy is the development of versatile shaped charges based on the outdated powders manufactured according to Dibasit producing technology.
Versalite chaped charges are designed for cutting metal structures, concrete, reinforced concrete, stone, wood and so on. They produce no splinters, and can cut items of different geometrical configuration. They can also be used for welding by shot of heterogeneous metals to produce composite materials of bimetallic character possessing original final properties. There are various sizes and forms of the YK3 * charges developed, including flexible and elastic ones. It is possible to face hollow cone with metal or metallopolymer. YK3 charges are the most profitable ones in comparison with the available analogs at the

cost of using in the composition of „outdated" powders (up to 98 %) as the cheapest raw material at the present time. The technical characteristics of YK3 charges are presented in Table 2.

* YK3 - versatile shaped charge

Table 2

TECHNICAL CHARACTERISTICS OF YK3 CHARGES

Charge type		Mass without facing, g / m		Mass with facing, g / m		Thickness of the obstacle being cutted, mm
YK3	1-01	1300	60	1800	100	12 - 14
YK3	1-02	2600	100	3200	110	15 - 17
YK3	1-03	4500	120	6000	140	20 - 25
YK3	2-01	274	20	355	20	7 - 10
YK3	2-02	480	30	640	30	9 - 11
YK3	3-01	75	10	110	10	3 - 4
YK3	3-02	120	10	160	15	6 - 8
YK3	3-03	800	40	1140	50	15 - 20
YK3	3-04	1160	50	1740	100	19 - 25
YK3	3-05	1800	100	2700	100	25 - 30
YK3	3-06	2600	130	3900	150	30 - 33

Notes: Recommended for the following use:

YK3 1-01, YK3 1-02, YK3 1-03, - for reinforced concrete, bricks, stones;

YK3 2-01, YK3 2-02, Yk3 3-01, YK3 3-02, YK3 3-03 - for cutting metal structures;

YK3 3-04 aned further - metal structures of large thickness.

YK3 charges represent the result of the research works, they belong to the industrial explosives of „C" class, group 1.

By the danger degree of treatment they belong to class 1, subclass 1.1, compatibility group „A". Serial number - OOH-0237.

According to the Decision of the GOSGORTEKHNADZOR of Russia HP 11-22/259 of 16.07.1992 the YK3 charges (Specifications 3-7509103.293-92) are allowed for regular use.

At the present time the NIIPM carries out an interesting research on development of the compositions based on the ballistite fuel (besides industrial explosives) aiming to obtain signal lights of five flame colours: red, green, yellow, light blue, white, so as to enlarge the civilian products assortment.

The components of ballistite fuel (nitro-ethers, stabilizers for chemical stability, catalysts of burning, technological additives) are used as oxidizers and burning-and-binding base.

The high energy alloy of metal is included in the composition with the aim to increase energetic content. A special polymer and colour-forming inorganic salts of different metals ensuring the required radiation spectrum are used for special luminous effects.

The basic composition is given in Table 3.

Table 3

Components	Percentage
1. Ballistite base	74 - 76
2. Alloy (metallic)	10 - 12
3. Special additives	14 - 16

The elements of signal cartridges are manufactured according to the ordinary technology of semi-continuous production of the ballistite fuel, it consists of the following steps: preparation of components, mixing, squeezing out, rolling, drying, moulding, cutting into segments.

This process of manufacture features high performance (10 thousand cartridges/hour) and the possibility to produce the elements of different size and geometric form.

A set of main characteristics is determined for the signal lights: technological, ballistic, explosive and special properties - light intensity, specific light emission, flame colour purity.

The composition possesses good technological properties, chemical stability like standard fuels, small sensitivity to mechanical and thermal influence; the composition and the products of combustion do not contain toxic substances harmful to human beings and the environment.

The main characteristics of signal lights composition are given in Table 4.

Table 4

Main characteristics	Figures				
	CK composn.	C3 composn.	C* composn.	CГ Composn.	CB composn.
1. Flame colour	purple-red	green	orange-yellow	light-blue	white-rose
2. Colour purity, P	99.9	72.5	70	70	70
3. Specific light emission Lo, cd	6552	1298.3	3214.8	14730	679.1
4. Light intensity, cd	5152	974	544.4	830.9	389
5. Burning rate at P=0, 1MPa, t=20 °C	1.81	1.82	1.25	2.08	1.64
6. Inflammation delay, msec at t=20 °C	35	34	38	34	24
7. Temperature of the intensive decomposition beginning, t idb, °C	140	140	145	135	140
8. Sample combustion time, sec	5.77	5.4	7.53	5.16	8.38

There are given some values (for comparison) of pyrotechnical compositions for various purposes:

compostions	light intensity, cd	colour purity, %
fireworks	360 - 10000	50 - 98
signal lights	1200 - 15000	not less than 70.

At the present time the specifications for the signal lights compositions are issued, in the laboratory the pilot lot of cartridges for signal rockets is manufactured: 120 thousand pcs of red, green, light blue colours.
The results of full-scale tests of signal cartridges:
cartridge combustibility: 95 to 100 %; time of combustion: 5 to 6 sec.; height of lifting: 50 to 70 m.
In 1994 we have received an order to produce 1.3 million cartridges.

Conclusions

New industrial explosives are developed: Dibasit according to TY* 3-7509103.325-93, YK3 charges according to TY 7509103.293-92, and the fuel for signal rockets according to TY 7509103.344-92. The process of production is mastered.
GOSGORTEKHNADZOR of Russia has admitted the new industrial explosives for use.
As opposed to the previously developed explosives the State Comittee of sanitary-epydemiological inspection has allowed Dibasit for manufacture, supply and use, sanitary certificate HP 232 of 17.12.1993.
The content of „outdated" powders (up to 98 %) in the composition of the new industrial explosives helps to solve the problem of utilization of ballistite compositions removed from the armament.
The process of the new civilian materials manufacture is close to the process of the ballistite compositions manufacture, due to this fact it is possible to employ trained highly qualified staff and to use the equipment becoming free because of military orders decreasing.
The scientific and production base allows to fulfill any orders of potential customers of the designed products.

* T Y - Specifications

BALLISTITE POWDER AS A MATERIAL SOURCE FOR POLYMER VARNISHES, ADHESIVES, AND COMPOSITES

V.V. Shilov

Institute for Macromolecular Chemistry, Ukraine's
National Academy of Sciences, Kiev

The basis of ballistite powders is nitrocellulose with a moderate degree of nitration (up to 12 per cent of nitrogen), mixed with low-molecular compounds which ensure the needed properties of powders. Physically, the latter play the role of solvents, plastificators, binders. Modern technologies involve a very wide application of nitrocellulose, as a rule, in mixtures with low- and high-molecular compounds rather than in pure form. Therefore, the most viable way of applying ballistite powders is their modification to parameters typical of traditional varnishes, adhesives, paints, and other composites. Possibilities for this lie in preparing thermodinamically compatible compositions of the nitrocellulose-solvent, nitrocellulose-resins, nitrocellulose-polymer types.

Considered are thermodynamic aspects of nitrocellulose interaction with various low- and high-molecular components, including filled systems based on peculiarities of concentration dependence of free energy of the systems under consideration, phase equilibrium curves and dynamics of phase transformations in them. Analyzed are viable ways of nitrocellulose compatibility with various high-molecular compounds, the systems' thermodynamic chances related to the removal of solvent.

It is shown that nitrocellulose may be the basis for real systems with coerced compatibility of components (of the interpenetrating networks and semi-interpenetrating networks types).

The analysis demonstrates possibilities of developing and manufacturing polymer varnishes for various applications, polymer coatings, and adhesives by way of composition modification of ballistite powders.

H. Krause (ed.),
Conversion Concepts for Commercial Applications and Disposal Technologies of Energetic Systems, 221.

POSSIBILITIES AND EXPERIENCE OF THE AMMUNITION EFFECTIVE UTILIZATION ON THE INDUSTRIAL OBJECTS

Uley G. Schukin, V. V. Galkin, Urey P. Nagebin

One of the most effectiv directions of the great volume ammunition realization is their utilization in mining and construction works blasting. Such utilization will allow to receive the greatest proceeds without any capital expenditures.

Among writing off ammunition suitable for utilization in mining and construction works there are ballistic hard propellants, powders and all kinds of the engineer weapons.

Special features, physical, explosional and exploitational properties of the ammunitions allow to apply them for all industrial blasting methods.

It will be illustrated by concrete examples.

The demining charges Y3 were successfully tested for presplitting and for lining initiation of blasthole charges.

The tests of Y3's linear initiators were carried out in open pit "Temir-Tau" at research-productional enterprise "Sibruda" in the 250 mm diameter blasthole.

It was established in a result of the experimental blasts:

1. A substitution of the standart concentrate boosters by linear Y3 for blasting granulated ANFO hole charges provided relative charge performance incrising by 1,7 time.
2. Crushing rock in the bench toe zone and rock blasting quality were considerably better when Y3 boosters are used.

During tests Y3 charges were compared with contour blasthole charges from ammonite 6KB (cartrige diameter 90 mm).

Experimental blasts had allowed to establish the following:

1. A labour-intensivity of Y3 charge gathering decrise in 3 time regarding to connection cartridges of commercial explosives.
2. High water-resistance of Y3 charges allowed to charge the contour holes for few days before a blast, and to improve all mining works systems.

Another variety of a military wares is a propellant intended for moving rockets and shells.

H. Krause (ed.),
Conversion Concepts for Commercial Applications and Disposal Technologies of Energetic Systems, 223–226.

It is known, that charge ability to crush and to move the hard rock is determined by its energetical and detonational characteristics such as specific explosive heat, absolute weight strenght (AWS), velocity of detonation (VOD), detonation front pressure, gas pressure in a hole and explosive strenght.

The carried out calculations had shown that over energetical and detonational characteristics hard propellants and powders are equal to commercial granulated TNT level.

To determine the propellants and the powders ability for charging blast holes in dry and watered rocks there were carried out the complextional researchs in the laboratory and rock massive.

On a base of tests in the laboratories and rock massives there were developped safe technologies and charge constructions for propellant and powder using in blasting rock under different mine-geological conditions.

The propellant and powder tests were carried out at the open-pits where blasting rock fulfils association "Sousevzrivprom", stock company "Transvzrivprom", firm "Hydrospecstroy" and so on open pits of firm "Roudprom" and others enterprises fulfiling commercial blasts.

Ballistic hard propellant and powder were used as blasthole charges. The blasted rocks were a limestone (strengch factor f=4-14), granite (f=12-16), dolomite (f=6-10) and quartzite (f=10-20). Hole diameter was 150-250 mm.

Blastholes were watered. The charges were in holes from 4 hours up to 5 days. The holes were hand-charged.

It was established:

1. The technology, developped by "Rosconversvzrivcentr" research institute, allow successfully to use the powder and the cast propellant elements for rock blasting at the complicated mine and geological conditions.

1.1The element of the jet and artillery systems may be used for great volumes rock blasting. For this purpose the most effective hole diameters were determined for each ammunition variety.

1.2The hard ballistic propellant elements with linear charge up to 3 kg/m may be used for contour blasting.

1.3The linear cumulative charges with regulated groove angle were developed to form mining face at the rocks with different physical and mechanical properties.

1.4To raise the rate of rock crushing was proposed the propellant charge disign which allows to regulat a specific bulk strength of the charge.

2. When hard propellant blasted in watered conditions, toxic oxide carbon concentration is decreased by 1,2-1.3 time with comparison dry charge blasting.

3. It need to fill in space between charge and hole wall with active oxidiser (for example with AN) to decrease oxide carbon concentration.

4. In partly watered blastholes it is economically and ecologycally advisably to use ballistic hard propellant and powder charges together with the most simple commercial explosives.

In accordance with preliminary estimational dates the most integrated industrial application the written off armament elements may be realized by means high velosity shock systems for hard environment destruction - bullets ad shells.

Traditional drilling and blasting methods for hard rock destruction have the whole range of the essential disadvantages, in spite of their wide exclusive application. So they are high labour- and energy- intensivity, cyclic operations, limited capacity (drifting velocity), people presence necessity and accordingly high injuring factor.

Crushing rock by means of high velocity shock of shells may be automated and carried out remotely without people presence near face. Moreover the crushing rock may be combined with loading and transporting loosen rock. In this case the drifting velocity may be considerably increased than one for traditional method and injurious effects (dust, noise and gas) may be localized by special arrangements. These dates are confirmed by US Bureau of Mines fulfilled experimental rock loosening by means of high velosity shock of inertial body. For this method a specific energy expenditure at granite composed 10.4 Dj/cm cub. against 36 Dj/cm cub. for drilling and blasting drifting method.

At average it may be taken that 100 mm calibre shot looses 0.5 m cub. of hard rock. For underground drifting of 1 km long working by section 10 m square it is necessary 10000 - 15000 shots. And approximately the same quantity of 20-50 mm calibre shots is necessary for working contouring.

The working unit may be consist of a shooting system, a barrel, a breech, a charging mechanism, a loading and transporting mechanisms.

Summing up we can state:

The Research Institute "Rosconversvzrivcentr" had developed the highly effective and economical rock blasting technology based on the written off ammunition application. This technology is inculcated successfuly on the objects of the firm "Rudprom", the assosiation "Sousvzrivprom", the joint stock companies "Transvzrivprom" and "Hydrospecstroy" and other enterprises also. There was already blasted about 500 t the demining charges, 1000 t hard ballistic propellant and 4000 t powder.

At this article it was shown a perspective application of slightly modernized ammunition for hard rock loosening and their come down at the hard reached working faces.

ABRASIVE WATER JET CUTTING OF MUNITIONS AND EXPLOSIVE INCINERATION IN A FLUIDISED BED

J. CAMPOS, R. FERNANDES, F. MATOS and A. PIRES
Laboratory of Energetics and Detonics
Mech. Eng. Dep. - Fac. of Sciences and Tech. - Univ. of Coimbra
3000 Coimbra
SPEL - Sociedade Portuguesa de Explosivos, S. A.
Sta. Marta de Corroios - 2840 Seixal
PORTUGAL

ABSTRACT. The end of the cold war has led about 5 million tonnes of surplus or obsolete munitions. These munitions must be disassembled or their casings must be opening up in order to re-cycle or incinerate their contents.
The application of an abrasive water jet system for the cutting of munitions (from 40 to 155 mm) is presented. The cutting process is influenced by factors as water pressure, kind and concentration of abrasive, angle of incidence of water jet and rotation speed. Any reaction of cast TNT explosive has been observed, our energetic material case study. The explosive-water slurry, formed during cutting operation, and all the other kind of explosives which can not be integrated like components in industrial civil explosives, must be incinerated. The gas composition can be predicted by THOR computer code. An original pressurised fluidised bed combustion system has been designed and built in order to incinerate them. It is composed by a pressurised fluidised bed, warmed by the combustion of a propane-air mixture as fluidising gas. A cyclone and an exhaust gas system close the combustion cycle. The exhaust gases are analysed by an IR spectroscopy gas analyser.

1. Introduction

The environmentally responsible disposal of used chemical products, the personnel safety, the prevention of sea dumping and the integration pollution control origin the developing of new concepts and studies of disposal technologies of energetic systems and materials. The potential classic strategies are based in store at a secure site, or stabilise and store, or to recover them in order of a future civil and military use. Their technological options can be classified in thermal destruction, chemical and electrochemical oxidation, chemical transformation and recently in biodegradation. (vd. Jupp (16), Demilitarisation Technology (9)).

H. Krause (ed.),
Conversion Concepts for Commercial Applications and Disposal Technologies of Energetic Systems, 227–245.

The end of the cold war has led about 5 million tonnes of surplus or obsolete munitions, most of them in Germany and Eastern Europe. The lack of adequate disposal facilities, safety, high cost of disposal and environmental considerations origin their demilitarisation. These munitions must be disassembled or their casings must be opening up in order to re-cycle or incinerate their contents. For environmental reasons re-cycling or conversion of explosives, propellants and other chemical components is highly desirable. This is a old problem that takes now new dimension as a consequence of the large quantities evolved and the existence of a recognition of our environmental complex problems.

The relatively modest portuguese armed forces, the insufficient quantities to stabilish a large national demilitarisation programme, the difficulties in locating suitable storage sites and the national policy of develop clean combustion and incineration technologies (vd. (3), (4), (22)) lead us to select two of all the possible disposal procedures of energetic systems:

- the access and segregation of propellant and explosive from its metallic confinement,
- the clean incineration of not possible recovered energetic materials.

The application of an abrasive water jet system for the cutting of munitions was selected to realise the first procedure, due to its qualities of safety and cost effectivity for the opening up munitions of diameters from 40 to 155 mm. It is qualified as explosion-free equipment in an explosive vapour or gas atmosphere. The cutting process is influenced by factors as water pressure, kind of abrasive, angle of incidence of water jet and rotation speed.

The explosive-water slurry, formed during cutting operation, and all the other kind of explosives which can not be integrated, as components, in industrial civil explosives, must be incinerated. There are many kind of energetic materials included in this case (generally single and double base propellants, TNT, other rocket propellants, black powders, metallic oxides) generating pollutants in their gaseous exhaust products (NOx, CO, HCl, HF, SO2, H2S). The mean values found in bibliography are presented in Table 1 (vd. Folk (10), van Ham (22)).

Table 1 - Toxic gases, in weight %, from combustion products of explosives (Folk(10)).

Explosive	NOx	SO2	HCl	H2S
Single base	10	-	-	-
Double base	19	-	-	-
TNT	0.01	-	-	-
Rocket prop	-	-	25	-
Black powder	-	0.1	-	6-8

The old most common way (now not allowed, illegal) to dispose of obsolete equipment with energetic substances has been to dump them in isolated areas, preferably into deep spots in the sea. Meanwhile, there is a certain inadequacy concerning the laws in the E.E.C. about the emission of polluting gases such as

NO_X, SO_X and CO and solid ashes. A reference of threshold values for gaseous exhaust products in The Netherlands (vd. van Ham (22)) can be seen in Table 2.

Table 2 - Threshold values for gaseous exhaust products in The Netherlands (van Ham (22)).

Product component	Threshold value [mg/m3]
NOx	70
CO	50
SOx	40
HCl	10
Heavy metals	1
Cd/Hg	0.05
PCD´s (dioxines)	0.1 nanogr. TEQ/m3
Total solid dust	5

The composition of incineration products is predicted by THOR computer code, calculating the equilibrium compositions of gas and solid species, for a constant pressure and temperature, as a function of its initial concentration in the explosive water slurry and as a function of the concentration of energetic material in air.
An original pressurised fluidised bed combustion system has been designed and built in order to incinerate them. It is composed by a fluidised bed, warmed by the combustion of a propane-air mixture as fluidising gas. The energetic materials (explosives or pyrotechnic compositions) are pre-mixed with silica sand particles and injected by a twin screw extruder. A cyclone and an exhaust gas system close the combustion cycle. An acquisition and safety data system control all the gas products of combustion.
In order to test and optimise the incineration equipment it has been selected only one typical energetic material - TNT - justified by its well defined chemical pure composition and very known properties.
The followed text presents, successively, the results of applying the abrasive water jet cutting equipment to a mortar grenade ammunition, the THOR code, applied to our energetic material case study (TNT), the temperature and pressure conditions of the pressurised fluidised bed combustion system connected to an exhaust gas analyser.

2. Water abrasive jet cutting

2.1. SELECTED TECHNOLOGY.

Mechanical cutting technologies have been long used in portuguese industry SPEL to disassembly munitions. In order to prevent "hot spots", that could lead to ignition, large volumes of coolant are necessary to circulate. These methods are unsuitable for the large scale, high volume operations and many different sensibility of explosives and propellants, required to deal with current stockpiles of munitions. Many studies are being carried out on alternative technologies (vd. Demilitarisation Technology (9)) like punching, shearing, cryofacture, abrasive water jet. This last method has been identified as potentially the best technology, due to its safety and minimal equipment exposure to potential explosions, and a great flexibility of cutting of a wide range of munitions with simple manipulation systems.

There are two basic types of abrasive water jet cutting systems:

- entrainment systems in which abrasive is conveyed by an air flow to the cutting head where it is entrained into a water jet moving at over 500 m/s,
- suspension systems with direct injection, generating an abrasive/water slurry upstream to the cutting head.

This second system is more reliable and safe, because it avoid any gas bubble inside stream and allow a very good control over the jet process, with maximum particle velocities close to the average velocity.

2.2. THE USED EQUIPMENT

The used equipment, supplied by GBI Cold Cut Systems, is qualified as explosion-free in an explosive vapour or gas atmosphere. The system is based on the second type of water jet cutting (vd. Figure 1) and it can fully enclosed in a 20 feet container.

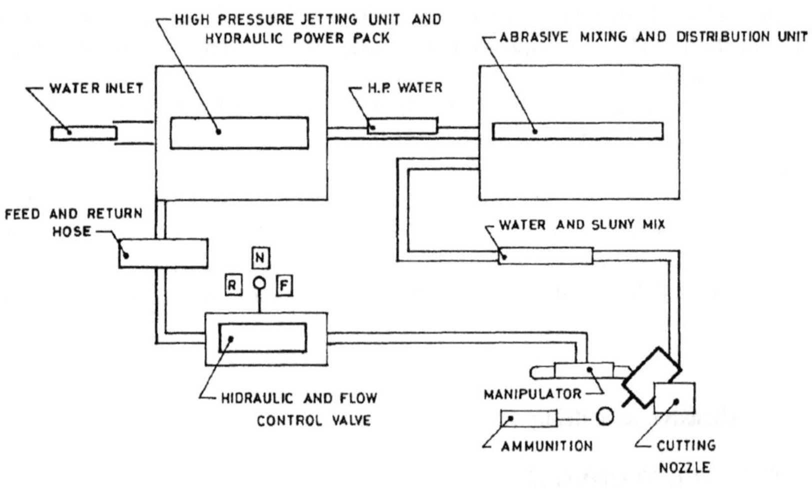

Figure 1. Abrasive water jet cutting functioning scheme.

Its main components (vd. Figures 2 and 3) are:

- an independent power engine, of 78 kW at 2500 rpm, which allows the location of this cutting systems in every clean disposal area;

- an high pressure water pump, with maximum working pressure of 370 bar and 73 liters of maximum water volume (under normal conditions it works at 250 bar and 50 liters);

- a hydraulic motor for robotic actuation of cutting head system, with a nominal and working pressure respectively of 50 and 30 bar, allowing no electric wires and circuits in working area;

- two robotic systems, providing the velocity and position of the cutting head, as a function of ammunition type and dimensions (vd. Figures 4 and 5);

- a cutting head with different diameter nozzle - normal it is used a 2.2 mm nozzle;

- several control instruments and remote controllers, including a safety shut down actuator system.

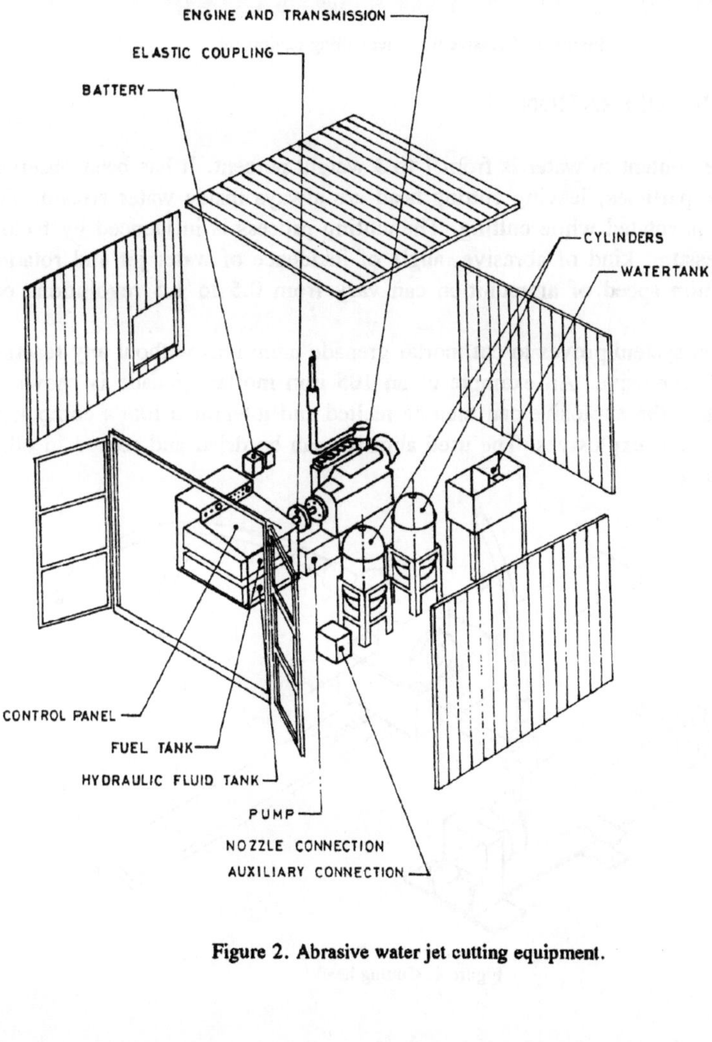

Figure 2. Abrasive water jet cutting equipment.

Figure 3. Abrasive water jet cutting equipment.

2.3. CUTTING OPERATION

The abrasive content in water is from 1 to 3 weight percent. It has been observed that abrasive particles, leaving cutting head are always inside water stream. The ammunition is rotated while cutting. The cutting process is influenced by factors as water pressure, kind of abrasive, angle of incidence of water jet and rotation speed. Rotation speed of ammunition can vary from 0.5 to 1.5 revolutions per minute.

The presented system proves to cut mortar grenade munitions without any reaction of cast TNT explosive. An example of an 105 mm mortar grenade is shown in Figure 5. Then the explosive can then be melted and integrated like a component of industrial civil explosives. The used abrasive can be dried and reused in other cutting operation.

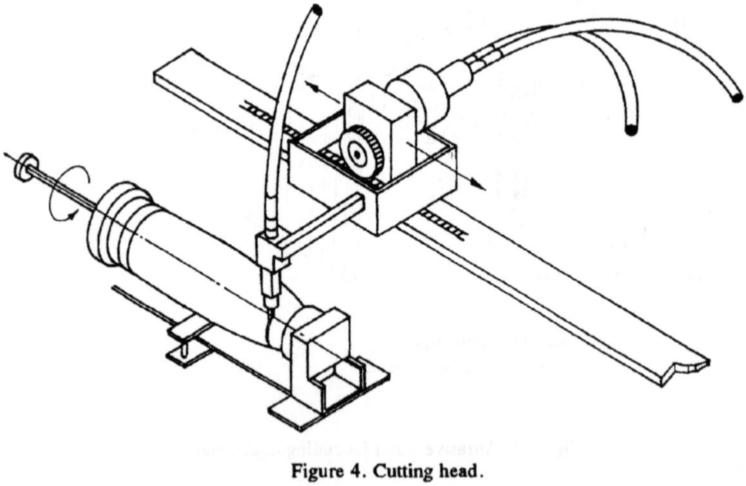

Figure 4. Cutting head.

Figure 5. Cutting operation.

2.4. CONCLUSION

The presented equipment proves to be:

- a safe and flexible cutting technology for munitions that also provides the maximum potential for recycling of explosives;
- possible to recycle water and abrasives contamined by explosives and chemicals;
- optimise parameters like safety, manipulation, environmental protection and disposal of wastes.

3. Incineration of explosive in a fluidised bed

3.1. SELECTION OF INCINERATION SYSTEM

There are some different ways of eliminating energetic substances, ranking from their storage in silos, their open air combustion or explosion. Aiming to solve this inadequacy the present work will try to develop a method for industrial application to reduce the amount of pollutants, which will benefit both physical and environmental conditions.

The method now proposed after longer discussion (vd. ICT Conference (22)) is the incineration of energetic substances using a fluidised silica sand bed, considered like one of the safest (vd. Michele et al. (8); Steward (20); Leithner et al. (17), Azevedo et al. (1) and van Ham (22)). In a fluidised bed due to the action of air flow, the particles of the bed float and act as a liquid. It allows the use of different kinds of combustibles particles in the same fluidised bed. Consequently the fluidised bed combustion has some advantages over other processes of combustion:

- security related to a better temperature homogeneity of bed;
- easy control of temperature, pressure and equivalence ratio of mixture;
- better heat transfer which means lower combustion temperatures and consequently a reduction of emission of polluting gases;
- higher density power involved;
- compensation of an increased flow of energetic substances owing to the pressurisation of the combustion chamber.

The great disadvantage of a fluidised bed combustion equipment comes not only from its ignition and security, during ignition period, but also from the necessity of keeping the fluidised bed warmed before introduction of energetic substances.

3.1.1. Selected incineration explosive case study - TNT - sand aqueous slurry.

The explosive-water slurry, formed during cutting operation, and all the other kind of explosives which can not be integrated like components in industrial civil explosives, must be incinerated. The selected explosive is TNT, dispersed in an aqueous slurry (vd. van Ham (22)). It is a very well known chemical defined explosive, with a carbon content larger than the stoicheiometry. It is also the main component of classical ammunitions. Consequently the theoretical predictions are based on water slurries of TNT, with 5%, 10%, 15% and 20% of water concentrations. The calibration and test mixtures are silica sand based mixtures, with a mass content from 5% to 10%.

The incinerator uses inert silica sand as the bed material having a size range of 21.9 µm to 564 µm (d_{50} = 377.7 µm) (vd. Figure 6).

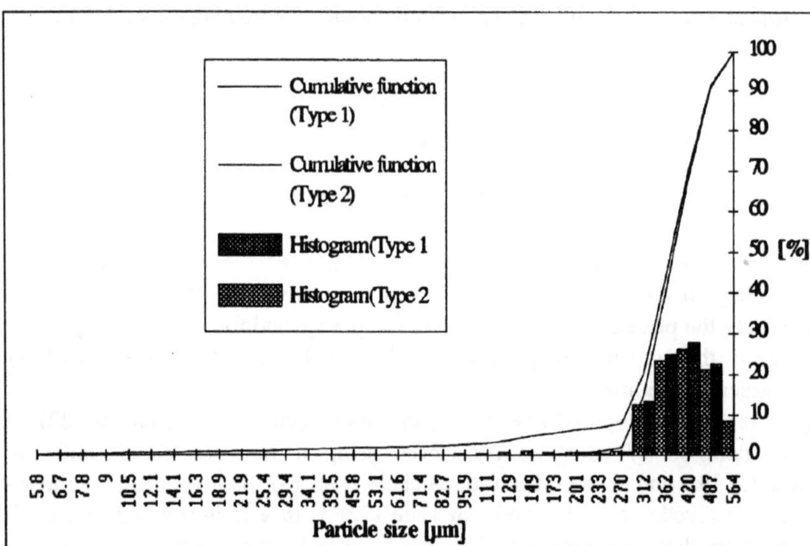

Figure 6. Cumulative curve of used two types of sand.

3.2. THEORETICAL GAS PRODUCTS COMPOSITION AND EVALUATION

The theoretical study starts with the prediction of combustion products using THOR code, based on theoretical work of Heuzé et al. (12), (13), and later modified (vd. IEPG Reports (14), Campos (3)) in order to calculate the composition and thermodynamic properties of explosive compositions, for isobar adiabatic combustion conditions. Several equations of state are used, namely BKW, Boltzmann, H9, H12 and JCZ3 (vd. Campos (3)). The results have been compared within themselves and with results of different codes in open literature (Quatuor Code from Heuzé et al. (12), (13), using the BKW EoS of Mader Code and TIGER Code (vd. Chaiken et al. (5)), and using the KHT EoS the Tanaka Code (21)).

3.2.1. Chemical equilibrium conditions.

The classical combustion system is generally a CHNO system. In our computer code it is possible to consider up to *m* atomic species ($m<20$) and to form *n* chemical components with these atomic species ($n<40$). Among these *n* chemical components, *m* are considered "basic" chemical components and *n-m* "non basic". The selection "ab initio" of the "basic" chemical components depends on the equivalence ratio *r* of the mixture, related to the stoicheiometry ($r=1$), and they are those which are expected to have significant concentrations in final products composition.

For a CHNO system it has been selected (vd. Manson (18); Heuzé (12)):

- CO_2, H_2O, O_2 and N_2 for poor mixtures ($r<1$),
- CO_2, H_2, H_2O and N_2 for rich mixtures ($r>1$) of initial low density and C(s), CO_2, H_2O and N_2 for rich mixtures of initial high density (initial condensed or solid components).

The mass balance yields a linear system involving *m* equations. In order to solve the problem it is necessary to add more (*n-m*) equations. These *n-m* equilibrium equations are determined by the method of Lagrange multipliers or the equilibrium constants (vd Campos (3)). Consequently the system of equations is formed by *m* linear mass balance equations and (*n-m*) non linear equilibrium equations.

In order to determine the chemical concentration of the *n* components, for imposed P and T conditions, two methods can been used:

- the chemical affinity method, proposed by Heuzé et al. (13);
- solving at first the system composed by the *m* "basic" components, and secondly adding one by one more components, optimising the relative concentration inside the group related to the same atomic species, for the minimum value of global Gibbs free energy $G = \Sigma\, x_i.\mu_i$, being the Gibbs free energy of each component $\mu_i = G_{oi}(T) + R.T.\ln P + R.T.\ln (x_i)$. The values of $G_{oi}(T)$ is the Gibbs free energy as a function of temperature. They can be obtained from JANAF Thermochemical Tables (15) and from polynomial expressions of Gordon and McBride (11). This second method is slower than the first, but avoid numeric problems (vd. Campos, 1991).

The solution of the composition problem involves simultaneously:

- the thermodynamic equilibrium, obtained with the mass and species balance, and the equilibrium condition $G = G_{min}(P,T,x_i)$, previously described, generally applying to the condensed phase the model proposed by Tanaka (21);
- the thermal equation of state (EoS);
- the energetic equation of state, related to the internal energy $E = \Sigma x_i e_i(T) + \Delta e$, $e_i(T)$ calculated from JANAF Thermochemical Tables (15) and from polynomial expressions of Gordon and McBride (11);
- the combustion regime, being $P_b = P_o$ constant for the isobar adiabatic combustion (equal final and initial total enthalpy $H_b = H_o$).

The selection of components are dependent of atomic initial composition. For a classical CHNO system it is considered the equilibrium compositions of CO_2, CO, H_2O, N_2, O_2, H_2, OH, NO, H, N, O, HCN, NH_3, NO_2, N_2O, CH_4 gases and two kinds of solid carbon (graphite and diamond). It is possible to include more species in final products composition, like S, SO, SO_2, using data from JANAF Tables (15) and polynomial expressions of Gordon and McBride (11). In order to evaluate the evolution of gas combustion composition, as a function of final combustion temperature, it is used to dilute gas products with a inert gas, generally N_2, and recalculate the composition for this new equilibrium conditions.

3.2.2. Theoretical results.

Calculated final gas products composition of TNT, as a function of dilution with air and N_2 representing respectively the isobar adiabatic combustion and isobar combustion, are show in Figures 7 and 8. There are no significant differences between EoS.

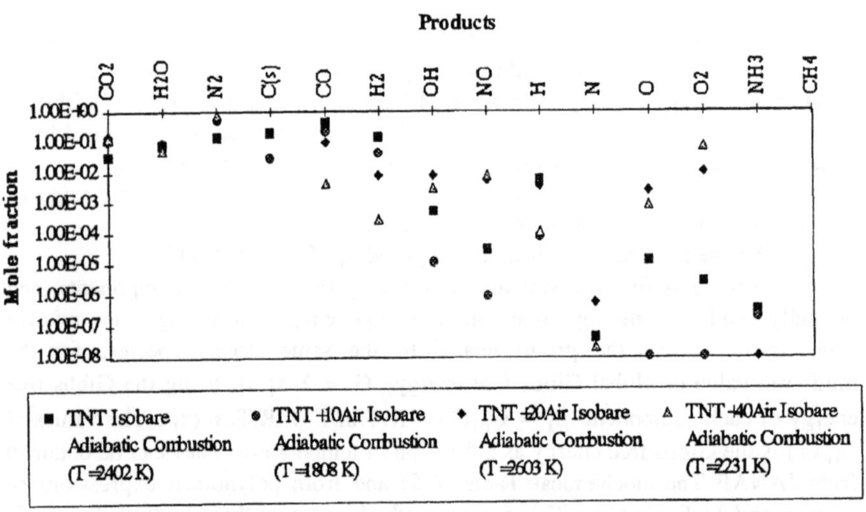

Figure 7. Gas and solids products composition of TNT as function of dilution with air.

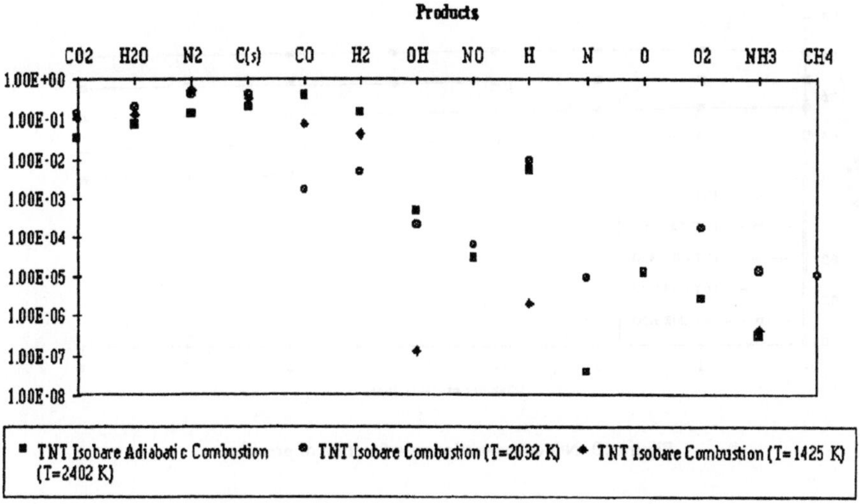

Figure 8. Gas composition of TNT as function of dilution with N_2.

Adiabatic combustion temperatures of water slurries of TNT and NO and CO concentrations, as function of initial chamber pressures, are respectively shown in Figures 9 to 11.

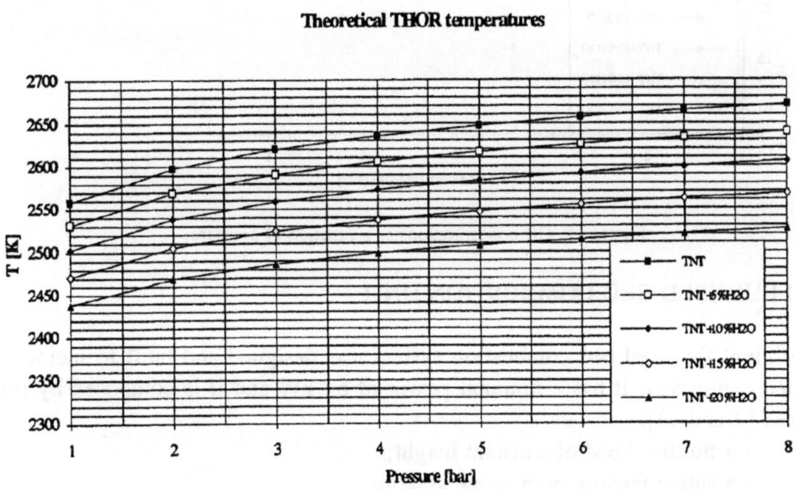

Figure 9. Combustion temperatures as a function of pressure.

Gases products composition, predicted using numerical code THOR, shows NO_x concentrations increasing and CO concentrations decreasing with low equivalence ratio related to stoicheiometry.

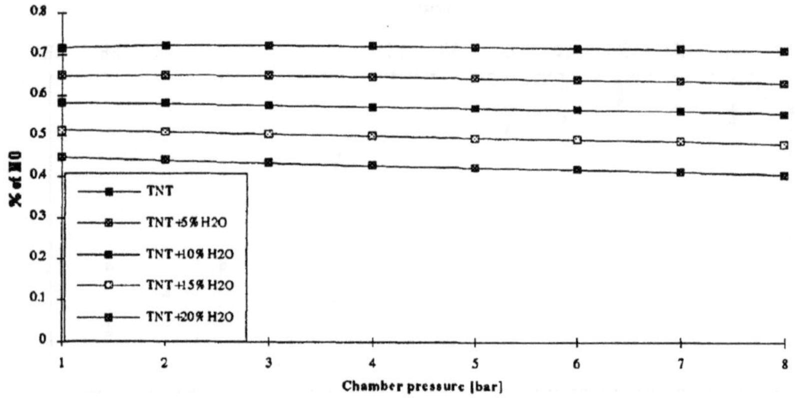

Figure 10. NO concentration as a function of pressure.

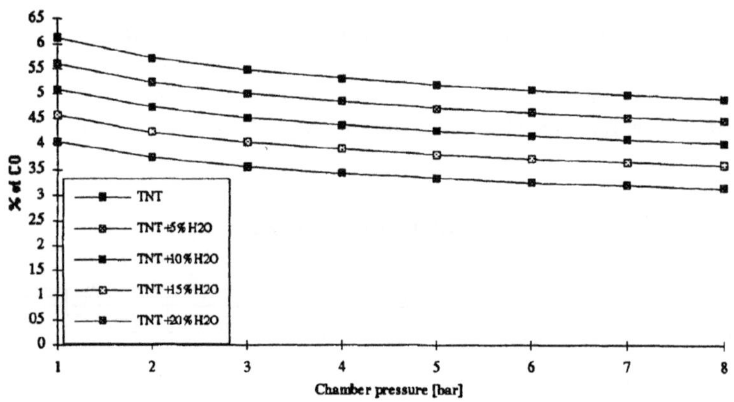

Figure 11. CO concentrations as a function of pressure.

3.3. FLUIDISED BED COMBUSTION UNIT

An original fluidised bed combustion system was designed and built to incinerate selected explosives. It has a nominal power of 61 kW and it is composed by (vd. Figures 12 and 13):

- a fluidised bed of constant height;
- a slurry feeding twin screw system;
- a pressure and temperature control with a real time data acquisition system;
- separation and filtration system for the gas combustion products and ash and solids products exhaust conducts;
- gas combustion analyser.

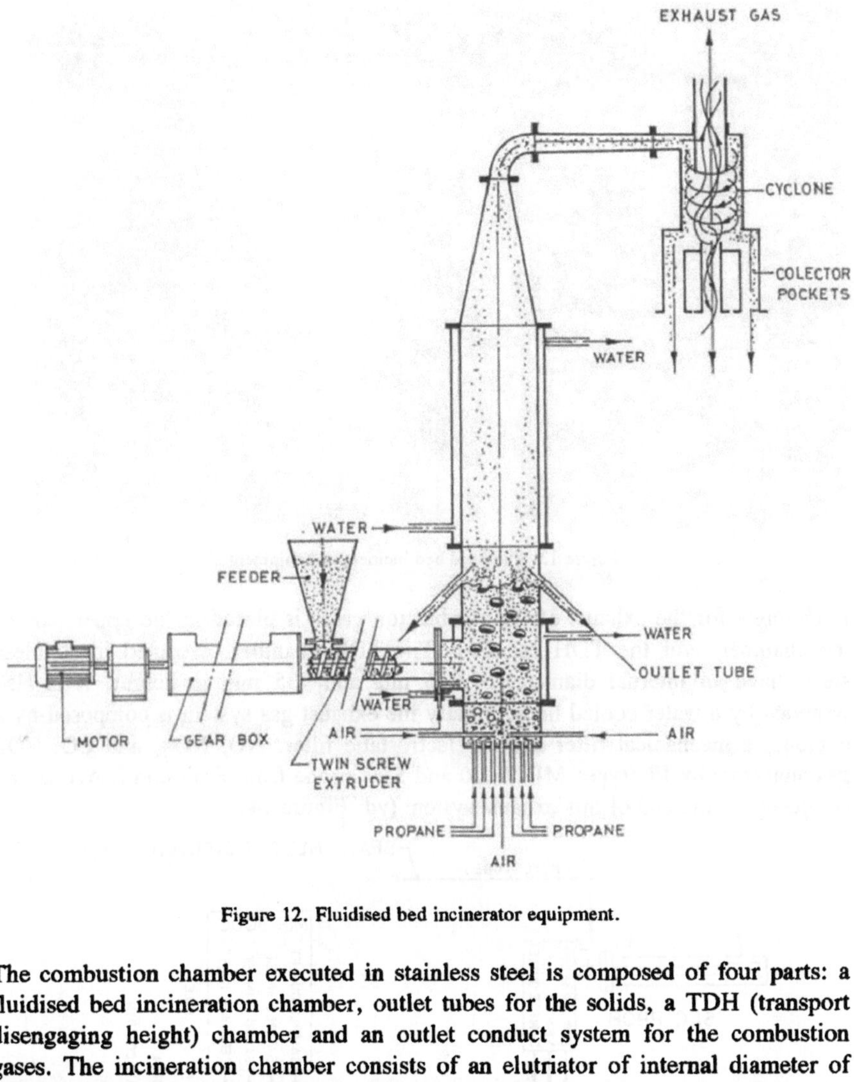

Figure 12. Fluidised bed incinerator equipment.

The combustion chamber executed in stainless steel is composed of four parts: a fluidised bed incineration chamber, outlet tubes for the solids, a TDH (transport disengaging height) chamber and an outlet conduct system for the combustion gases. The incineration chamber consists of an elutriator of internal diameter of 266 mm and 500 mm in height, which is wrapped by a water cooled liner. The fluidised bed combustion system is permanently warmed by the injection of a propane-air mixture as fluidising gas. Eight tubes placed on the base of the chamber allow the propane diffusion in air, conducted by an other system of eight tubes peripherally and one in the center. This distributed system ensure a final homogeneous gas mixture (vd. Figure 13).
Four exhaust tubes, for the sand and solids products, keeps constant the height of fluidised bed. These discharge tubes can be linked to the slurry feed system in a close circuit.

Figure 13. Fluidised bed incinerator equipment.

A chimney for the exhaust of the combustion gases is placed in the upper part of the chamber over the TDH chamber. The TDH chamber, executed in stainless steel, have an internal diameter of 266 mm and 735 mm in height. It is also wrapped by a water cooled liner. Finally the exhaust gas system is composed by a cyclone, a mechanical filter and an electrostatic filter. NO, NOx, and CO, CO2 gas analysers by IR (types MIR 9000 and SEC probe from Emission S.A.) can be connected to the end of this exhaust system (vd. Figure 14).

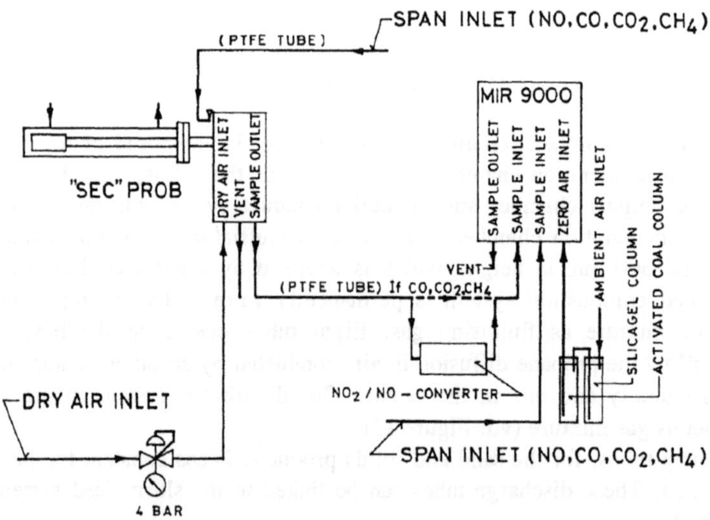

Figure 14. Gas analysers scheme by IR spectroscopy.

The energetic material feeding system is formed by a twin screw extruder for solids and two sonic flow meters for gases. To avoid the formation of slurry dunes in the connection tube, between the extruder outlet and the incineration chamber inlet, an entry of air was connected to the part of extruder outlet (vd. Figure 13). Two sonic flow meters ensure the rigorous control of air and propane flows (vd. Figure 15).

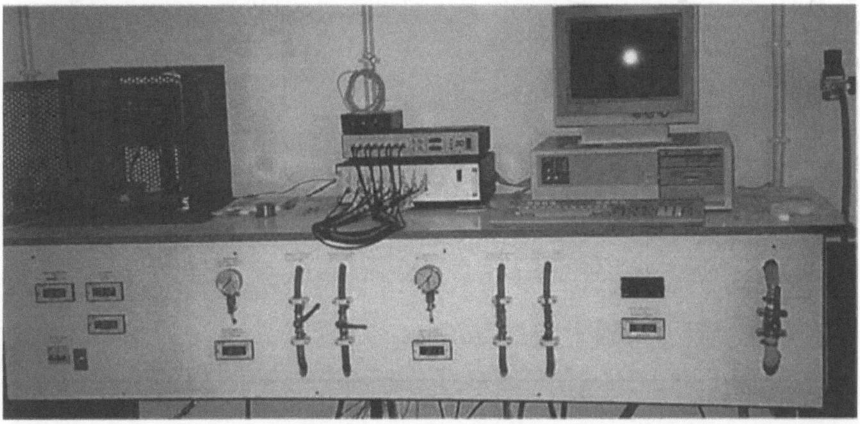

Figure 15. Control and safety systems.

The control and safety systems of the incinerator are composed (vd. Figure 15) by:
- a real time data acquisition temperature system, composed by a thermocouple amplifier, a analogue digital converter (3D equipment) and a personal computer;
- velocity control of the extruder changing rotating speed of driving motor;
- ignition device (vd. Figure 13) composed by a vortice propane/air flame;
- water cooling system;
- opening outlet tubes.

The design values and calibration curves has been presented in a previous work (vd. Pires and Campos (19)). The theoretical velocity for used mean particle of 350 mm is about 7m/s and the theoretical superficial gas velocity is about 0.3 m/s. Consequently the theoretical bubble velocity of 3 m/s is greater than the superficial gas velocity - this fact leads to bubble bursting at the surface of the bed and particles returning to bed.

3.4. EXPERIMENTAL RESULTS

Experimental bubble diameter, at the surface of bed, has been measured to be between 0.12 and 0.15 m, shows a good agreement with the theoretical values (d ≈ 0.18 m).

The experimental combustion conditions are shown table 3.

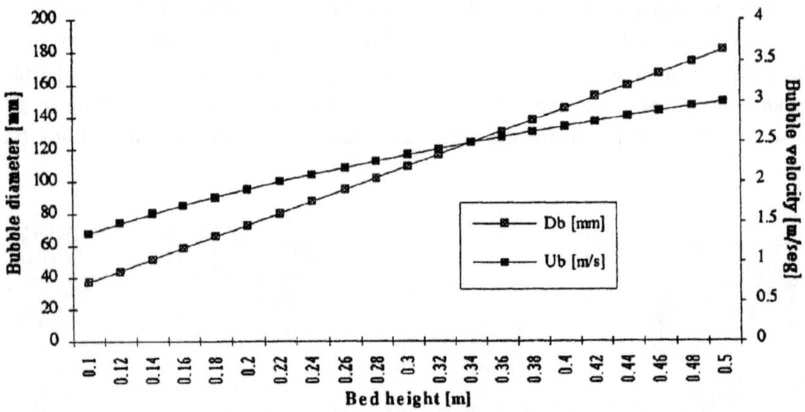

Figure 16. Bubble diameter and velocity as a function of bed height (theoretical values).

Table 3. Experimental combustion conditions.

Static bed height, H_{mf} [mm]	500
Diameter of the bed[mm]	266
TDH chamber [mm]	735
Average silica sand size, d_{50} [μm]	377.7
Theoretical minimum fluidisation velocity, U_{mf} [m/s]	7

There are two phases in a normal operation - the transient warming regime at the beginning and a permanent regime during its effective work.

In the transient regime the aim is to heat the sand bed. The combustion of the mixture of the fluidising gas (air and C_3H_8) heats the sand bed while the extruder is continuously sending clean sand into the chamber and the refrigeration circuit protects the chamber from any possible overheating.

Optimising conditions leads to pressures in the sonic flow meters approximately at 4.0 bar and 1.0 bar at 20 °C respectively for air and propane. Throughout all tests the sand bed was initially heated above 120° C. The freeboard temperature was kept between 700 °C and 900 °C (vd. Figure 17).

When nominal designed temperatures are attained, inside the chamber, the extruder gradually begins to send a mixture of sand and energetic material, while the flowing of C_3H_8 is being gradually reduced. It is necessary to take a special care with slurry mass flow feed in order to avoid the formation of big aggregates inside the fluidised bed. During the experiment the sand fills the chamber and are expelled through the discharge tubes (vd. Figure 13). Then the energetic material feeds the combustion chamber and the permanent regime is attained.

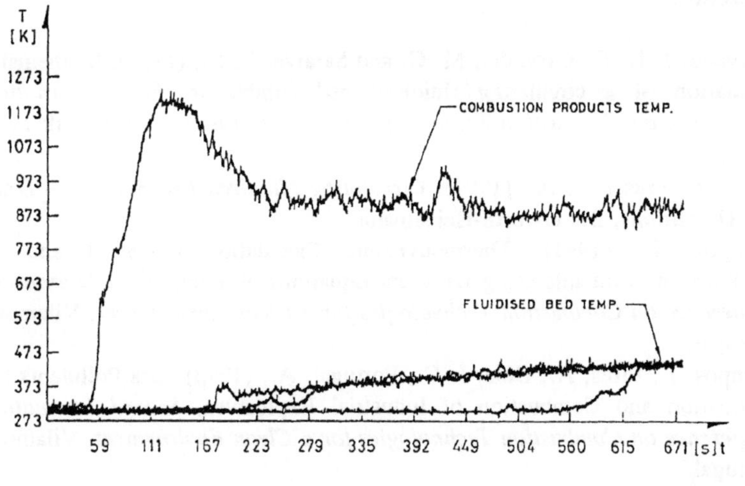

Figure 17. Fluidised bed combustion temperature as a function of time during pre-heating.

When the burning of energetic material begins the temperature inside the chamber rises. It shows that the combustion is taking place. When the incineration is finished we must feed the combustion chamber with clean sand for preventing pack aggregated slurry.

The experiments were done with air/propane mixtures of equivalence ratio to stoicheiometry from 0.701 to 2.099. The fluidisation temperature of the bed for a stable regime was considered best for the equivalence ratio 2.074 (3.9 bar $\leq P_{AIR} \leq$ 4.5 bar; 0.5 bar $\leq P_{PROPANE} \leq$ 1.0 bar), when the temperature fluctuation is a relative minimum.

3.5. CONCLUSIONS

An original fluidised bed combustion system for energetic materials is presented. It is composed by a fluidised bed, warmed by the combustion of a propane air mixture as fluidising gas. The energetic materials are pre-mixed with silica sand particles and injected by a twin screw extruder. A cyclone and a exhaust gas system close the combustion equipment. An acquisition and safety data system controls all the incinerating system.

There are two regimes of incineration: the warming transient regime and constant incineration regime. Obtain real temperatures are stable near 1200 K and its minimum fluctuations are detected for equivalence ratio to stoicheiometry near 2.

The energetic material case study has been TNT-sand aqueous slurry. This fluidised bed system proves to be a safe and programmable incineration system for explosives and propellants. In the future will be tested other energetic materials.

4. References

1. Azevedo, J. L. T., Carvalho, M. G. and Saraiva, P. C., (1991). Mathematical simulation of a circulating fluidised bed combustor. *First International Conference on Combustion Technologies for a Clean Environment*, Vilamoura, Portugal.
2. Basu, P., Fraser, S. A., (1991). *Circulating Fluidized Bed Boilers — Design and Operations*, Butterworth-Heinemann.
3. Campos, J., (1991). Thermodynamic Calculation of Solid and Gas Combustion Pollutants using Different Equations of State, *First International Conference on Combustion Technologies for a Clean Environment*, Vilamoura, Portugal.
4. Campos, J., Pires, A., Gois, J. C., Portugal, A., (1993). Gas Pollutants from Detonation and Combustion of Industrial Explosives, *Second International Conference on Combustion Technologies for a Clean Environment*, Vilamoura, Portugal.
5. Chaiken, R. F., Cook, E. B., and Ruhe, T. C., (1975). *Toxic Fumes from Explosives: Ammonium Nitrate-Fuel Oil Mixtures*. Report of Investigation n° 7867 - Pittsburgh Mining and Safety Research Center, Pittsburgh, Pa., USA.
6. Davidson, J. F., Harrison, D., (1971). *Fluidization*, Cambridge, Academic Press Inc., U.K. .
7. Davidson, J. F., Clift, R., Harrison, D., (1985). *Fluidization*, Cambridge, Academic Press Inc., 2nd Edition, U. K. .
8. De Michele, G., Grazzini, G. C., Miccio, M., and Tonfoni, C., (1991). Pilot and Laboratory Scale Fluidised Bed Combustors: A Comparison on Design, Operation and Results, *First International Conference on Combustion Technologies for a Clean Environment*, Vilamoura, Portugal.
9. Demilitarisation Technology, (1991). *International Seminar on Demilitarisation Technology for Explosives and Explosive Ordnance*, Royal Military College of Science and Directorate of Land Services Ammunition, Oxfordshire, U.K. .
10. Folk, F., (1990). Analysis of reaction products of propellants and high explosives. *Chemistry and physics of energetic materials*, Nato ASI Series, Kluwer Academic Press.
11. Gordon, S., McBride, B. J., (1971). *Computer Program for Calculation of Complex Chemical Equilibrium Compositions, Rocket Perfomance Incident and Reflected Shocks and Chapman-Jouguet Detonations*. Report NASA SP 273, NASA Lewis Research Center.
12. Heuzé, O., (1989). *Cálculo Numérico das Propriedades das Misturas Gasosas em Equilíbrio Termodinâmico*. Universidade de Coimbra, Portugal.
13. Heuzé, O., Presles, H. N., Bauer, P., (1985). Computation of Chemical Equilibria. *J. Chem. Phys.*, n°83 (9), pp.4734-4735.

14.IEPG - Portuguese Research Group - Campos, J., Luz, P. and Marques, C., (1989, 1990). *BKW - KHT - H9 Equation of State Calculations*, Report of Technological Area 25, Collaborative Technical Programme 1, Portugal.

15.JANAF, (1971). *Thermochemical Tables* - 2nd Edition. National Bureau of Standards, Washington DC.

16.Jupp, M.F. , (1991). An assessment of the european demilitarization scene. *International Seminar on Demilitarization Technology for Explosives and Explosive Ordnance,* Royal Military College of Science and Directorate of Land Services Ammunition, Oxfordshire, U.K. .

17.Leithner, R., Wang, J., (1991). One Dimensional Fluidised Bed Furnace Model, *First International Conference on Combustion Technologies for a Clean Environment*, Vilamoura, Portugal.

18.Manson, N., (1976). *Cours de Thermodynamique des Hautes Temperatures.* Ecole Nationale Superiore de Mecanique et d'Aerotechnique, Université de Poitiers, France.

19.Pires, A. and Campos, J., (1993). Clean combustion of energetic materials in a fluidised bed. *Second International Conference on Combustion Technologies for a Clean Environment*, Lisboa , Portugal.

20.Steward, F. R., (1991). Mathematical Model of a Circulating Fluidised Bed with Chemical Reaction, *First International Conference on Combustion Technologies for a Clean Environment*, Vilamoura, Portugal.

21.Tanaka, K., (1983). *Detonations Properties of Condensed Explosives Computed Using the Kihara-Hikita-Tanaka Equation of State.* Report from National Chemical Laboratory for Industry, Ibaraki, Japan.

22.Van Ham, N.H.A., (1991). Environmentally Acceptable Disposal of Ammunition and Explosives, *22nd International Annual Conference of ICT* - Combustion and Reaction Kinetics, Karlsruhe, Federal Republic of Germany.

PROCESSING OF ENERGETIC MATERIALS WITH SUPERCRITICAL CO_2

P. Gerber, G. Bunte, M. Niehaus, U. Teipel, H. Krause

Fraunhofer-Institut für Chemische Technologie (ICT)
P.O.Box 12 40, 76318 Pfinztal, FRG

Abstract

Because of their specific physical properties supercritical fluids show a specific solubility for various organic substances. Especially carbon dioxide with it's moderate critical data (T_c = 31,3 °C and p_c = 7,28 MPa), noninflammability and chemical inertness was found to be an attractive alternative to hazardous organic solvents.

Main objective of this work was to study potential applications on supercritical CO_2 in the processing of energetic materials. Generally, the solubility or extractabillity of energetic materials like TNT, RDX, HMX, PETN and NTO strongly depends on pressure, temperature and specific modifying agents. Presented topics and potential applications of SC-CO_2 are:

- examination of phase-equilibria-data for different concentrations of TNT in liquid and SC-CO_2 under static conditions using a special designed sample cell;
- study of extraction efficiencies for TNT-contaminated soil samples as a function of the used extraction conditions and especially as a function of the soil matrices (SFE in analytical scale);
- screening of the extractabillity of different explosives in SC-CO_2 as a function of the used extraction parameters (pressure, temperature, time, usage of modifiers) using a dynamical SFE-apparatus in analytical scale ;
- examination of usable processing conditions for the production of explosives like TNT and RDX with small particle sizes.

H. Krause (ed.),
Conversion Concepts for Commercial Applications and Disposal Technologies of Energetic Systems, 247–258.

1. Introduction

The use of supercritical fluids like SC-CO_2 inspite of conventional organic solvents becomes more and more attractive. Because of their specific physical and chemical properties supercritical fluids show a specific solubility for various organic substances. Especially carbon dioxide with it's moderate critical data (T_c = 31,3 °C and p_c = 7,28 MPa), noninflammability and chemical inertness was found to be an attractive alternative to hazardous organic solvents.

There are several industrial and/or analytical applications of SC-CO_2 as an environmental friendly and potential extraction and/or processing medium for example in the field of:

- production of flauvour compounds;
- extraction of pollutants from contaminated waste waters or soils;
- regeneration of activated charcol;
- extraction and fractionation of polymers;
- production of fine particles.

Main objective of this work was to study potential applications on supercritical CO_2 in the processing of energetic materials. Generally, the solubility or extractabillity of energetic materials like TNT, RDX, HMX, PETN and NTO strongly depends on pressure, temperature and specific modifying agents. Presented topics and potential applications of SC-CO_2 are:

- examination of phase-equilibria-data for different concentrations of TNT in liquid and SC-CO_2 under static conditions using a special designed sample cell;
- study of extraction efficiencies for TNT-contaminated soil samples as a function of the used extraction conditions and especially as a function of the soil matrices (SFE in analytical scale);
- screening of the extractabillity of different explosives in SC-CO_2 as a function of the used extraction parameters (pressure, temperature, time, usage of modifiers) using a dynamical SFE-apparatus in analytical scale;
- examination of usable processing conditions for the production of explosives like TNT and RDX with small particle sizes.

2. Potentential applications of SC-CO_2 in the processing of energetic materials

2.1 Investigation of the phase-equilibrium-system TNT - SC-CO_2

The phase equilibrium of the binary system trinitrotoluene (TNT)-carbon dioxide (CO_2) was under investigation. The homogeneous phase transition was determined during the heating-up-period of the system trinitrotoluene - carbon dioxide. The homogeneous phase transition is the point where the two phases are just completely soluble.

By the use of the synthetic method, the homogeneous phase transition of the system was determined in consideration of the following parameters:

$$0{,}5\ \text{wt\%} < c_{TNT} < 5{,}2\ \text{wt\%}$$

$$30°C < \vartheta < 140°C$$

$$150\ \text{bar} < p < 530\ \text{bar}$$

Figure 1 shows the phase equilibrium apparatus. To start the examination, a cylindrical vessel (figure 1) is filled with TNT and closed with high pressure bolted connection. In the vessel, two bores were used to measure the temperature and to fill in carbon dioxide. The vessel is heated with an electric band heater. Inside of one cover, a saphir glass is enclosed. The light of a halogen lamp passes through the saphir glass of the vessel and is reflected inside from the end wall of the piston.

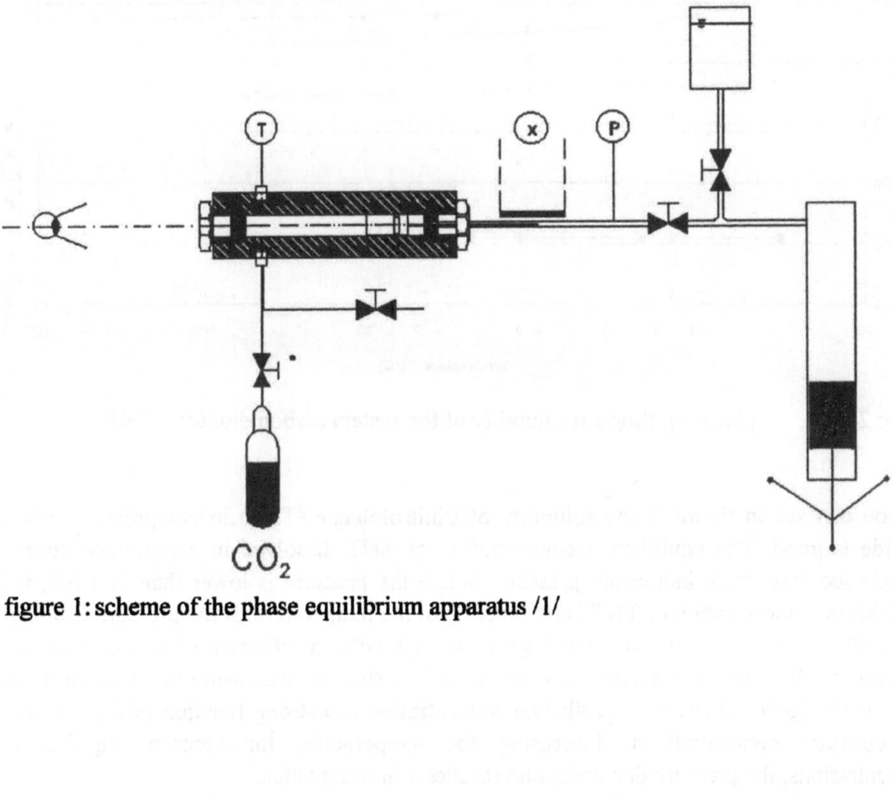

figure 1: scheme of the phase equilibrium apparatus /1/

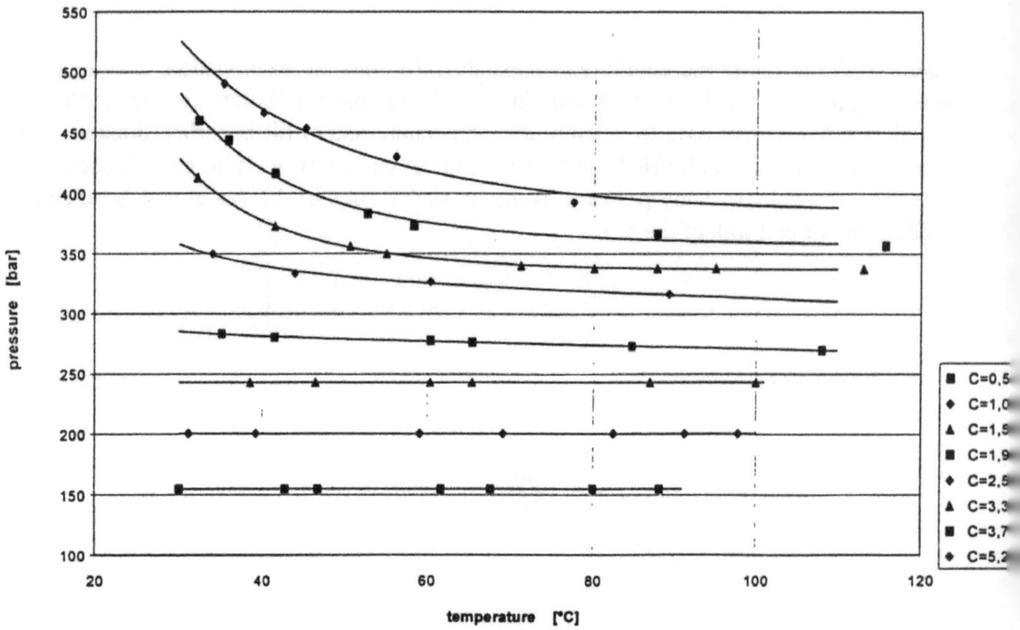

figure 2: phase equilibrium solubility of the system carbon dioxide - TNT

As you can see in figure 2, the solubility of trinitrotoluene (TNT) in compressed carbon dioxide is good. The equilibrium concentration of TNT dissolved in compressed carbon dioxide increases with increasing pressure. While the pressure is lower than 240 bar, the equilibrium concentration of TNT is no function of the temperature. If the pressure is higher than 240 bar or the concentration is higher than 1,5 wt%, a influence of temperature and pressure on the equilibrium concentration of TNT is studied. Increasing the pressure from 240 bar to higher values, the equilibrium concentration is a strong function of the pressure for constant concentrations. Increasing the temperature, for constant equilibrium concentrations, the pressure decreases and reaches a limiting value.

2.2 Extraction of TNT-contaminated soils - effects of matrices

The method of supercritical fluid extraction (SFE) with SC-CO_2 offers an attractive because rapid and solvent-saving alternative to the common used sample preparation techniques (e.g. soxhlet extraction). In this part of the study basic principles are presented to apply this technique in the field of evaluation and decontamination of soils contaminated with explosives like TNT. The objective of this work was to find suitable extraction conditions (e.g. extraction pressure, temperature and time) for a rapid and complete extraction of TNT-contaminated soils to get extracts which could be analysed directly with chromatographic and/or spectroscopic methods. Furthermore the influence of different soil matrices on the extraction efficiency was examined. Different model and also real soils were analysed to get soil-specific data / parameters and extracted with SC-CO_2 under various conditions.

Extractions were carried out in a dynamic extractor (SFE 703, Dionex) in which up to eight extraction cells of an internal volume of 2.5 ml were used. The modell soil samples (sand and different clays and schluff samples) were contaminated mechanically with 1 % TNT. For a sandy soil which has the highest particle sizes the extraction efficiency of TNT was examined as a function of extraction pressure. This extraction parameters is very sensitive as it could be seen in figure 3.

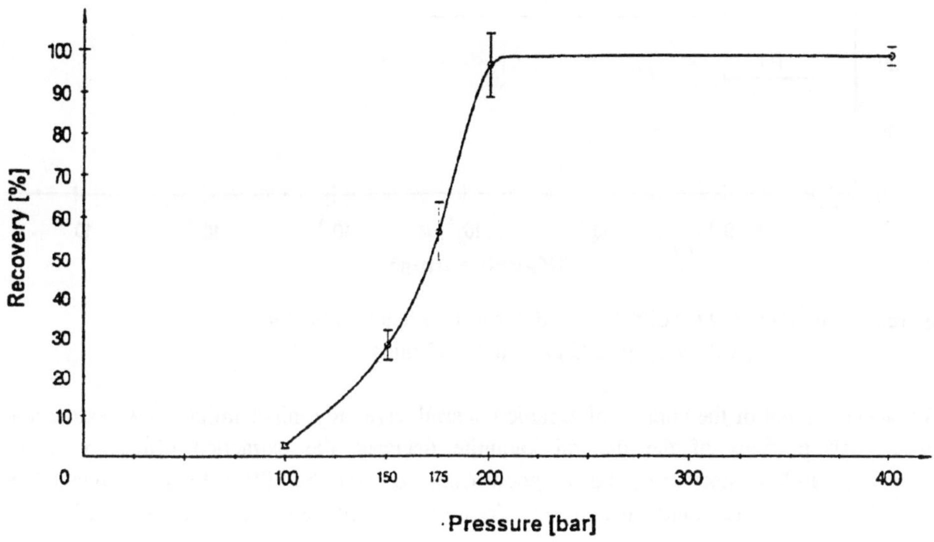

figure 3: TNT-recovery for a sandy model soil as a function of extraction pressure

Under conditions which lead to about 50 % of TNT-recovery (175 bar, 80 °C and 45 min) for the mentioned sandy soil sample, extraction experiments were carried out also for the other soils. The measured recoveries were always smaller than 50 % thus indicating that the soil matrix strongly influences the extraction efficiency of TNT (under these not optimized conditions).

Most important are the particle size distribution and the permeability (k_f-value) of the sample. Under constant SFE-conditions higher portions of bigger soil particle sizes lead to higher extraction rates. The same trend was found for higher k_f-values (see figure 4).

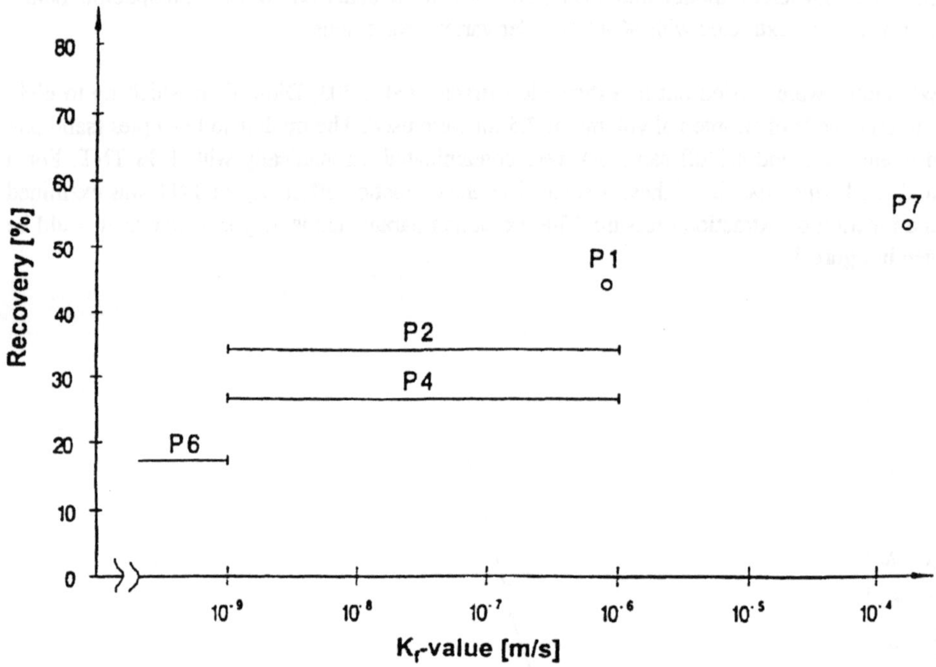

figure 4: TNT-recovery for different model soils as a function of their k_f-values (SFE: 175 bar, 80 °C, 45 min)

The water content or the content of organic material have only minor influence on extraction rates. High portions of the mineral kaolinite decrease the extraction efficiency. First experiments in laboratory scale showed good results to make the SFE-technique also suitable for sanitating explosive contaminated soils. For all soil samples SFE-conditions of 300 bar, 80 °C and 45 minutes lead to complete extraction.

2.3 Solubility of several explosives in pure and modified SC-CO_2

For e.g. the production of fine explosive particles by the Rapid Expansion of Supercritical Solutions (RESS) or the Gas Anti Solvent Process (GAS) investi-gations of the solubility of the explosive compounds in pure and modified SC-CO_2 are highly required.
The dissolving capacity of SC-CO_2 depends on solvent density and could be easily varied by changing the pressure. Especially for polar components the solubility could be raised using small amounts of organic solvents like acetone, methanol, ethanol and higher alcohols as a modifying agent.

Objective of this part of the study was the examination of the solubility of the explosives PETN, NIGU, NTO and RDX in SC-CO_2 as a function of the concentrations of the used different organic modifiers. To ensure supercritical conditions experiments were carried out with modifier amounts up to 4 %. The experiments were carried out in the dynamic extractor mentioned above but was extended by a modifier modul (SFE 703 with modifier modul, Dionex). Extraction conditions used were 300 bar, 50 °C for 30 minutes. Measured average CO_2 flows at the end of the extractor were 12 liters at a temperature of 28 °C and a pressure of 1 bar (this means 0,81 ml/min CO_2 at extraction conditions). The extracted amounts were quantified by HPLC-analyses to calculate the socalled recovery, R which is defined as the ratio of registrated mass of the extracted explosive to the total mass used for the experiment.

Figure 5 shows the recovery of RDX as a function of the modifier concentration. Without co-solvens, recovery of RDX is about 0,8 % thus indicating very low solubility in SC-CO_2. Except ethanol all used modifiers (addition of 2 mol% or 4 mol%) slightly enhance the dissolving capacity of the extraction fluid. Nevertheless large enhancements only occur with cyclohexanone and moreover with acetonitrile modified SC-CO_2.

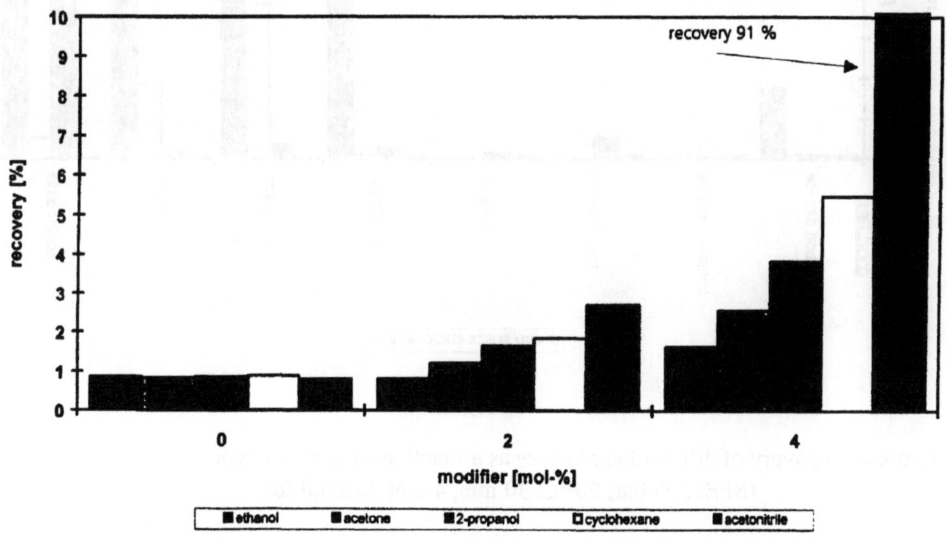

[GB1]

figure 5: Influence of modifier-concentration on recovery of RDX

Further extraction experiments were carried out for the explosives NIGU, RDX, NTO and PETN using pure SC-CO_2 as well as seven different organic modifiers (in each case 4 mol %). Measured recovery rates are shown in figure 6. With pure SC-CO_2, only PETN was extracted. The recovery of the other samples was below 1 % indicating no mutual solubilities in pure carbon dioxide.

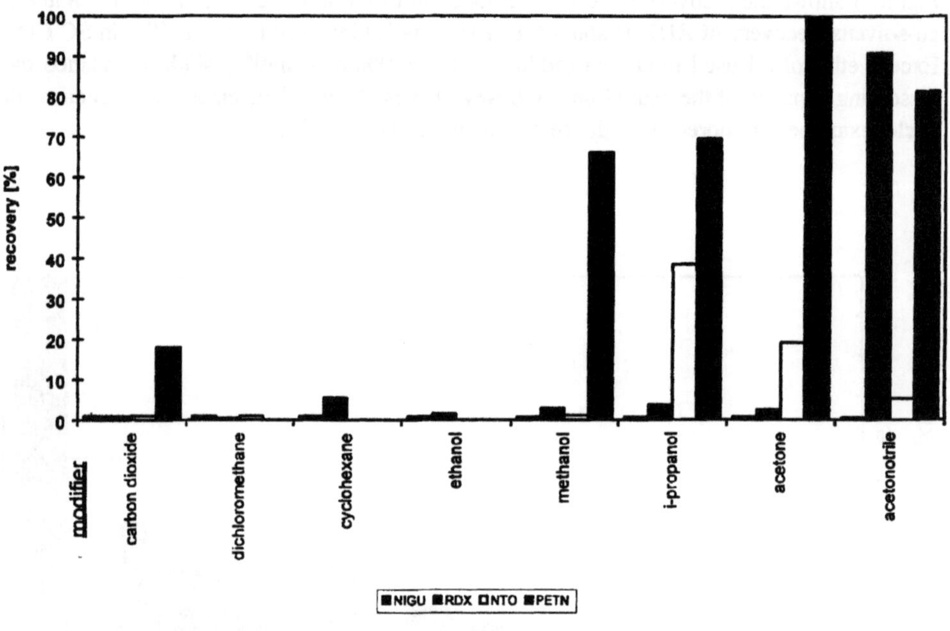

figure 6: Recovery of different explosives as a function of modifier-type
(SFE: 300 bar, 50 °C, 30 min, 4 mol-% modifier)

Besides the results for RDX mentioned above we found that especially for NTO and PETN the extractability could be raised drastically using acetone and acetonitrile and in minor degrees also with i-propanol. Under the extraction conditions mentioned above a mixture of 4 mol% i-Propanol with SC-CO_2 extracts NTO up to 38 %. For the explosive PETN recovery rates of 18 % were measured for pure carbon dioxide. 4 mol % of acetone lead to 100 % recovery, using 4 mol % of i-propanol or methanol yielded 66 to 70 % recovery. For NIGU none of the studied modifiers lead to acceptable solubilities.

From the experiments it is clear to be seen that there is a need of choosing the right modifier for a single explosive by empirical measurements. There is no easy correlation possible e.g. of the dipole moment of the used organic co-solvent with the solubility of the different explosives. Results indicate, that the solubility of PETN in pure SC-CO_2 is high enough for the production of fine particles with the RESS-Process. It seems to be possible to produce fine RDX and NTO particles with acetonitrile or i-propanol as a modifying co-solvent.

2.4 Production of fine explosive particles

When a solution existing of a compressed gas and a dissolved explosive is rapidly depressurized, the dissolved solid becomes insoluble in the low pressure gas. The rapid expansion of supercritical solution (RESS) exploits this phenomena. During the process, a loaded supercritical fluid is expanded through a nozzle creating high supersaturation in the jet. Fast nucleation and growth of the crystalline particle occurs. If carbon dioxide is used as a supercritical fluid the RESS process offers advantages over conventional size reduction methods. Heat sensitive and solvent free products can be achieved with this process.

The rapid expansion of supercritical solutions (RESS) was explored by several authors as a novel route to the formation of microparticles. Ohgaki [2] produces fine stigmasterin particles by the rapid expansion of a supercritical CO_2-solution. Amorphous fine particle and whisker-like crystals (0,05 - 3 μm) were obtained with different pre-expansion pressures. Johnston [3] obtained submicron particles from different polymers. Loth [4] described the mirconisation of phenacetin with supercritical fluids.

An influence of the process parameters, i.e. concentration, pre- and post-expansion pressure and pre- and post-expansion temperature and geometry of the nozzle on the particle size distribution was not studied at all. Merely Mohamed [5] examined the influence of concentration, pre- and post-expansion pressure and temperature of the system carbon dioxide - naphthalene. No particle size distribution was measured, only the size of the smallest and biggest particles were measured. For these investigations, trinitrotoluene was used as a model substance, while the solubility of trinitrotoluene in carbon dioxide is well known.

The experiments were carried out wiht a pilot plant which was built in cooperation with SITEC AG (Switzerland). The extractions are possible up to 80°C and 300 bar. The maximum massflow of carbon dioxide is limited to 20 kg CO_2/h. Figure 7 shows the sheme of the pilot plant.

Liquid carbon dioxide (purity 99,95 Vol %) is undercooled (W2) to avoid cavitation in the membran pump (P). After the compression to pre-expansion pressure, the fluid is heated to the extraction temperature (W3). The supercritical fluid loaded with anthracene leaves the extractor (V = 0,6 l). With an additional heat exchanger (W4), the solution is heated to pre-expansion temperature. In the separation vessel, the supercritical solution is expanded through a nozzle. The expanded gas will be condensed (W1) and recompressed or let off. After the experiment, the separation vessel is opened and the particles are collected. The particle size is measured by laser diffraction spectroscopy (Malvern Master Sizer X).

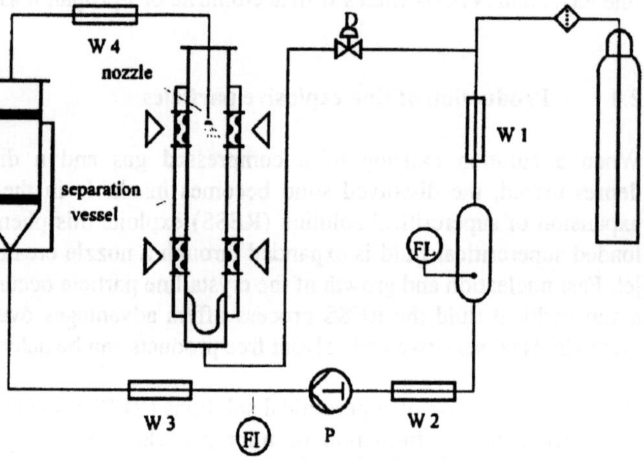

W: heat echanger
FL: liquid CO_2 - reservoir
P: jetting pump
FI: determination of mass flow
D: pressure control

figure 7: scheme of the experimental apparatus

The extractor was filled wiht 3 g trinitrotoluol (TNT) and 300 g $MgSO_4$. The particle size of TNT was determined with laser diffraction spectroscopy to 54 μm. Preliminery investigations show that it is possible to dissolve TNT completely from a $MgSO_4$ matrix. A typical particle size distribution for trinitrotoluene (TNT), which is received by the RESS - process, is shown in figure 8.

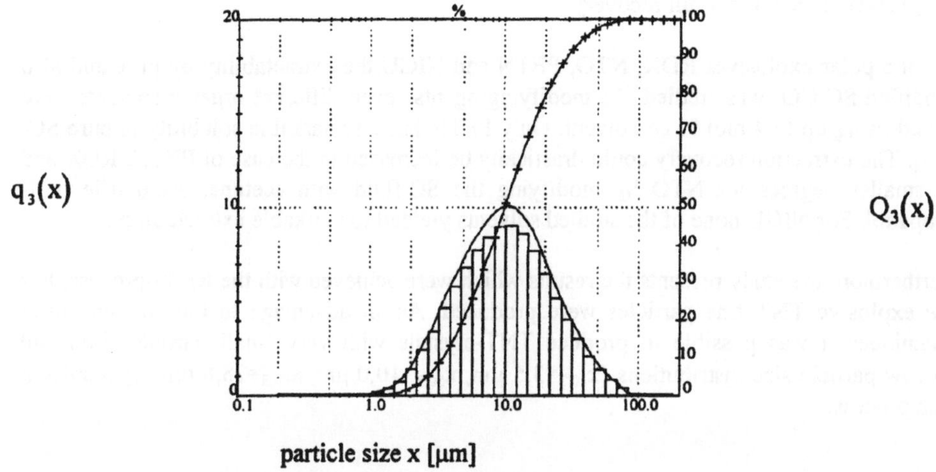

result: $x_{10,3}$ = 3,5 μm, $x_{50,3}$ = 10,0 μm, $x_{90,3}$ = 28,8 μm

figure 8: volume sum distribution Q_3 (x) and volume density distribution q_3 (x)of trinitrotoluene,

For these investigations, the process parameters of the RESS-process are:

- pre-expansion-temperature T = 115°C
- pre-expansion-pressure p = 220 bar
- post-expansion pressure p = 1 bar
- nozzle diameter d = 100 μm
- mass flow m_{CO2} = 6,5 kg/h
- temperature - extraction T = 75 °C

It is possible to process different explosives, if the solubility in carbon dioxide is higher than 1 %wt. While an explosive is not soluble in carbon dioxide, a organic solvent could be additionally added to process this explosive with the RESS - process.

3. Summary

From the results of the mentioned studies it is clear to be seen that the supercritical fluid carbon diopxide is a potential extraction or processing medium. It could be used alternatively to common hazardous organic solvents.

The phase equilibria data were determined for the system SC-CO_2 - TNT. Between pressures of 150 bar to 530 bar the phase equilibria concentrations were examined and found to be a function of pressure and temperature.

For explosive-contaminated soils which mainly contain TNT and it's destruction products pure SC-CO_2 was found to be a very good extraction medium. For experiments under deliberate non optimized extraction conditions it was noticed that the soil matrix especially the particle size distribution and the related kf-value of the soils influences the extraction efficiency of TNT. In order to get higher extraction recoveries mainly the pressure of SC-CO_2 is well suited to get full recovery.

For the polar explosives RDX, NTO, PETN and NIGU the extractability by pure and also modified SC-CO_2 was studied. As modifying agents seven different organic solvents were tested using up to 4 mol % co-solvent. Only PETN has a remarkable solubility in pure SC-CO_2. The extraction recovery could drastically be increased in the case of PETN, RDX and in smaller degrees for NTO by modifying the SC-fluid with acetone, acetontrile or i-propanol. For NIGU none of the studied solvents yielded to markable extractability.

Furthermore the study presents the results which were achieved with the RESS-process. For the explosive TNT fine particles were produced. As an advantage to the common used techniques it was possible to produce TNT-crystals with very small particle sizes and narrow particle size distributions ($x_{10,3}$=3,5 µm, $x_{50,3}$=10,0 µm, $x_{90,3}$=28,8 µm) by a solvent free process.

4. Literature

/1/ (Article in Book) Merz L., Hirth T., Michelfelder B.
Extraction of polymer binders from powder injection molding workpieces with supercritical carbon dioxide, 3rd International Symposium on Supercritical Fluids, 17.-19. 10. 1994, Strasbourg (France)

/2/ (Article in Journal) Ohgaki K., Kobayashi H., Katayama T., Hirokawa N
Whisker formation from jet of supercritical fluid solution, J. of Supercriticl Fluids, 3 (1990), 103-107

/3/ (Article in Journal) Matson S., Johnston K. P., Combes J. R., DeSimone J. M., Formation of Poly (1,1,2,2 -tetrahydroperfluororodecylacry-late) submicron fibers from particles from supercritical carbon dioxide solutions, ACS, Macromolecules 1995, 28, 3182 - 3191

/4/ (Article in Journal) Loth H., Hemgesberg E., Properties and dissolution of drugs micronized by crystallization from supercritical gases, International Journal of Pharmaceutics 32, 1986, 2-3

/5/ (Article in Journal) Mohamed R. S., Halverson D. S., Debenedetti P. G., Prud'homme R. K., Solids formation after the expansion of super-critical mixtures, SCS Symp. Series 406, Am. Chem. Soc. (1989 b) 355 - 378.

DISPOSAL OF ENERGETIC MATERIALS BY ALKALINE PRESSURE HYDROLYSIS AND COMBINED TECHNIQUES

G. Bunte, H. Krause

Fraunhofer-Institut für Chemische Technologie (ICT)
Joseph-von-Fraunhofer-Str. 7, D-76327 Pfinztal, FRG

Abstract

Due to the reduction of armament and especially due to the german reunification we are met by objective the disposal of energetic materials.
In the next few years there will be approximately 40,000 t of energetic materials to dispose off. Environmentally friendly disposal methods available for the different propellants, explosives and pyrotechnics are urgently needed. The main component of gun and rocket propellants is the energetic polymer nitrocellulose. One method to dispose off nitrocellulose containing propellants is the combination of rapid chemical destruction by pressure hydrolysis and the biological degradation of the reaction mixture. The study describes the results of pressure hydrolysis of different gun and rocket propellants. Under alkaline conditions (propellant to NaOH ratio 2,3:1; reaction temperature 150 °C; pressure below 30 bar) biological degradable reaction products were formed. The main products in the liquid phase were simple mono and dicarboxylic acids. Dependent on the reaction conditions 30-50 % of the nitrogen content of the propellants was transformed to nitrite and nitrate. The gaseous nitrogen containing products were N_2 (16-46 %), N_2O (2-23) %, NO_x (0-5) %. Overall 40-60 % of the propellant nitrogen was transformed to gaseous products. In the solid residues a nitrogen content between 2 and 9 % was found. The residues were mostly due to additives used in propellant manufacturing. In the case of nitrocellulose pressure hydrolysis below 30 bar and reaction temperature about 150 °C is sufficient.

H. Krause (ed.),
Conversion Concepts for Commercial Applications and Disposal Technologies of Energetic Systems, 259–269.

1. Introduction

Old ammunition found in the stocks of the former east german army (NVA=Nationale Volksarmee) amount to approximately 300,000 t. The amount of energetic material contained here is estimated at approximately 40,000 t. The energetic materials consist of gun and rocket propellants, high explosives and pyrotechnic compositions. The main component of gun and rocket propellants is the energetic polymer nitrocellulose. The high explosives consist mainly of TNT and the nitramines RDX and HMX. The composition of the pyrotechnic ammunition is very complex. For example NVA-ammunition to form artificial smoke was manufactured from Lindan residues that means a mixture of hexachlorocyclohexanes with aluminia. The disposal of these HCH compositions is a serious hazardous waste problem. The objective is to make specific environmentally friendly disposal methodes available for the different groups of energetic materials and pyrotechnic compositions.

The disposal processes must be

- safe
- technically feasible
- environmentally safe (in respect to soil, water, air) and
- economic in operation.

The disposal processes are dependent on the composition and thus on the chemical properties of the different materials involved. From the view of quantity the disposal of nitrocellulose (NC) based materials (25,000 t of gun and rocket propellants is of highest priority. The most common method to dispose off energetic materials is incineration. Different incineration systems either rotary kiln or fluidized bed incinerators are in use. On incineration of energetic materials high amounts of air pollutants are evolved. Especially the emission of NO_x is a serious problem which requires additional dinox stages to meet the german legal emission limits. Therefore incineration of energetic materials needs the cost extensive investment of fluid gas purification.

Furthermore, incineration plants don't have the acceptance of general public independent of technical questions. Hence, alternative disposal methods for energetic materials are highly requested. One way to dispose off nitrocellulose based propellants is the combination of chemical and biological disposal. This concept means to destroy the explosive character of the energetic material by a simple and cheap chemical reaction producing a biodegradable reaction mixture to study the feasibility of this technic.

It is known that nitrocellulose can be destroyed by alkaline hydrolysis /1, 2/. The application of this method to complex gun and rocket propellants has to be tested. Preliminary experiments have shown that alkaline hydrolysis of gun propellants under normal pressure conditions needs serveral hours to decompose the propellant. Whereas alkaline hydrolysis (under pressure, below 20 bar) shows comparible results with a reaction temperature of about 150°C in shorter times by a factor of 8-10. To study the feasibility of pressure hydrolysis as a chemical step in the chemical/biological disposal of NC-based energetic materials we investigated the pressure hydrolysis of 5 realistic gun and rocket propellants /3/. Hereby we have to characterize the reaction products to assess the biodegradability of the reaction mixture and also the overall pollutant emission especially in comparison to incineration technics.

2. Experimental

2.1 Materials

Five gun and rocket propellants were studied. The compositions are given in table 1.

Table 1 Composition of gun and rocket propellants

Name	P 1	P 2	P 3	P 4	P 5
NC[1]	97	80	57,5	57	56
NG[2]	-	14	26,7	28	26
DNT[3]	-	2	8,5	11	9
Centralite	-	2	2,9	3	2,9
DPA[4]	2	-	-	-	-
Graphite	1	1	-	-	-
Potassium salt	-	1	-	-	-
Lead	-	-	1,4	-	-
Dibutylphthalate	-	-	-	-	4,5
Vaseline	-	-	1	1	1

[1] NC = nitrocellulose

[2] NG = nitroglycerin

[3] DNT = 2,4-dinitrotoluene

[4] DPA = diphenylamine

The propellants were of different size in the form of sticks and cylinders. To get comparable results we granted the propellant down to a grain size of 1-5 mm. P 1 is a typical single based propellant the other are so called double-base propellants. In case of P 5 we found changing amounts of TNT as substitute of DNT.

2.2 Pressure Hydrolysis Equipment

The pressure hydrolysis equipment (figure 1) consists mainly of an 180 cm^3 autoclave (o_d = 70 mm, i_d = 48 mm) in combination with manometer, pressure release value and sampling value. Pressure and temperature (thermocouple) are simultaneously registrated during the reaction period. The reaction vessel is heated by an electrical resistance heating which allows heating rates of 10 °C/min.

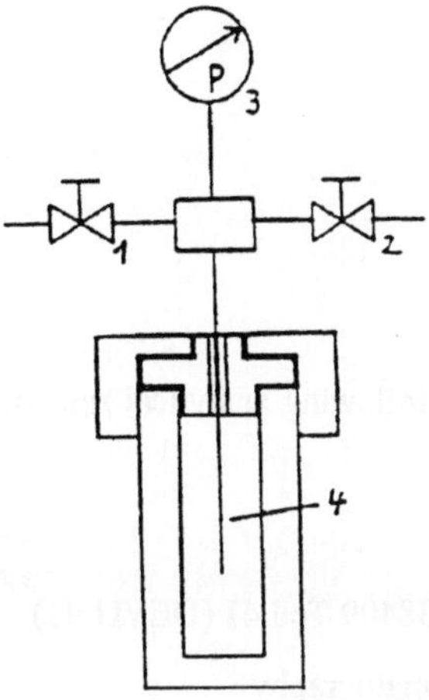

1 pressure release value

2 sampling value

3 manometer

4 thermocouple

Fig. 1 Pressure hydrolysis equipment

2.3 Hydrolysis conditions

The starting conditions of the reaction were P:H_2O 1:10, P:NaOH 2,3:1. 120 ml of the reaction mixture were put in the autoclave. The autoclave was closed and evacuated and afterwards filled with helium. This was necessary to determine exactly the molecular nitrogen which was formed during hydrolysis. The reaction vessel was heated up to 150 °C (15 min). This temperature was held over a period of 1 hour. The exothermic hydrolysis leads to small temperature rises in the magnitude of 10 °C which was compensated by a reduced heating. During the experiments the reaction temperature could be kept constant within ± 3 °C. After the reaction period of 1 hour the reaction mixture was cooled down to 90 °C and separated into gas phase, liquid and solid residues.

2.4 Characterisation of Reaction Products

2.4.1 Aqueous Solution

To characterize the composition of the aqueous solutions following analytical methods were used:

- determination of chemical oxygen demand (COD) DIN 38409 Teil 41 (DEVH 41);
- determination of nitrite and nitrate by ion exchange chromatography (IC in combination with conductivity detector, Fa. GAT);
- determination of short chain carboxylic acids by ion exclusion chromatography (IC with conductivity detector, Fa. GAT)
- PH value by glass electrodes ;
- infrared spectra of the liquid phase using an ATR-Circle Cell in combination with FTIR-spectroscopy (Circle Cell with ZnSe-crystal, Fa. Spectra Tech Inc.);
- IR-spectra of the residue of the evaporated liquid phase (KBr peletts).

2.4.2 Gas Phase

The gaseous phases were analyzed by:

- gaschromatography with thermal conductivity detector (TCD) with molecular sieve 13X and Poropak Q columns (FISONS-GC6000 VEGA);
- NO_x measurement by chemiluminescense technic (Modell 59A Fa. Rosemount)

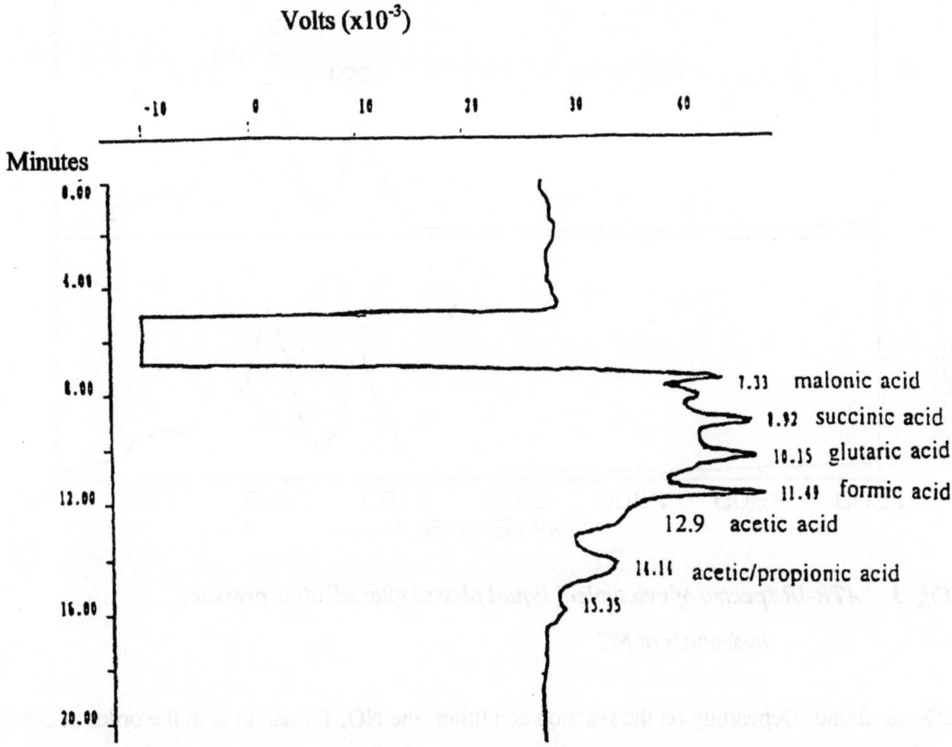

Fig. 2 ***Ionexclusion chromatogramm of the liquid phase after alkaline pressure hydrolysis of NC***

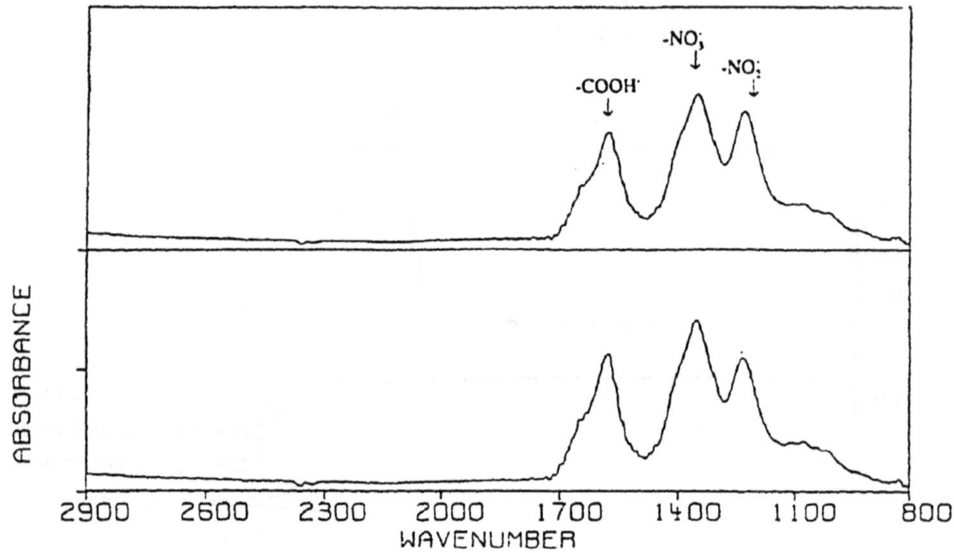

Fig. 3 ATR-IR spectra of two typical liquid phases after alkaline pressure hydrolysis of NC

CO_2 are found. Depending on the reaction conditions the NO_x formation is in the order of 0-5 %. The N conversion ratios of double base propellant P 2 are summarized in figure 4. It shows that nearly 40 % of nitrogen is converted to the harmless molecular nitrogen. Laughing gas is in amount of 10 %. Laughing gas and NO_x as air pollutants can be introduced in the subsequent biological stages where the denitrification takes place. The solid residues depend on the composition of the propellant:

- P 1: 2-3 % residue consisting of DPA and graphite
- P 2: 3-12 % with centralite, DNT was not found in the residue
- P 3: about 10 % were lead oxide Pb_2O_3 and centralite
- P 4: 5-16 % residue under enrichment of centralite
- P 5: 10 % residue with enrichment of centralite and dibutylphthalate.

Under the conditions of pressure hydrolysis diphenylamine as well as centralite as additives in propellants remain in the solid residue. The enrichment of those compounds in the residues is an advantage of the process because both compounds should not be transfered to the savage

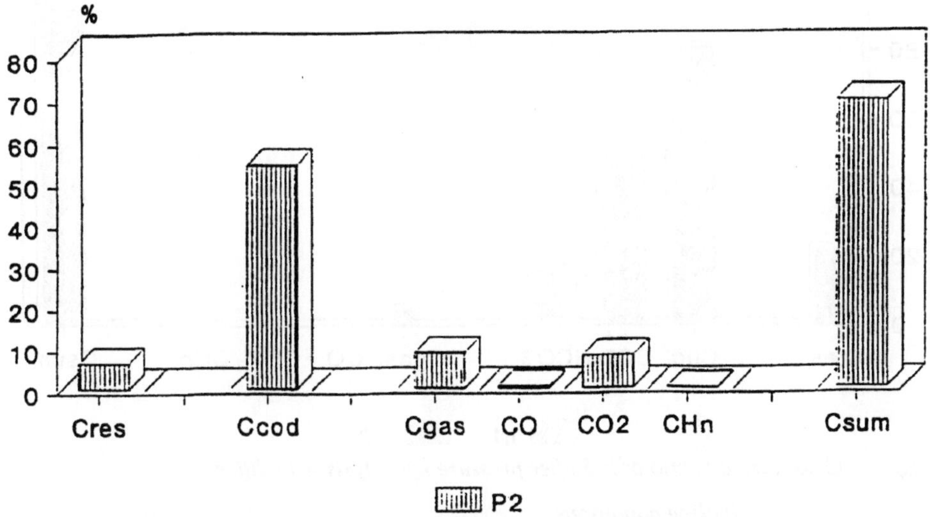

Fig. 5 C conversion ratio of P 2 after alkaline pressure hydrolysis

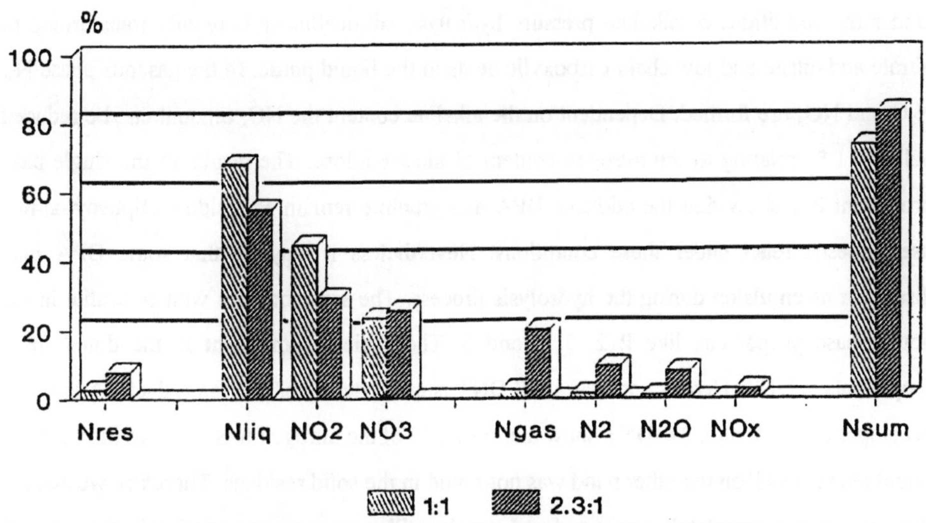

Fig. 6 N conversion ratio of P 5 after pressure hydrolysis with different alkaline conditions

Figure 7 shows the carbon conversion ratios with lower and stronger alkaline content. With stronger alkaline conditions the C-content in the residues is reduced (from 21,9 % to 9,4 %). The COD value is significant higher (87 % instead of 71 %) whereas the gas production (mainly CO_2) is drastically reduced (from 4,5 % to 0,01 %).

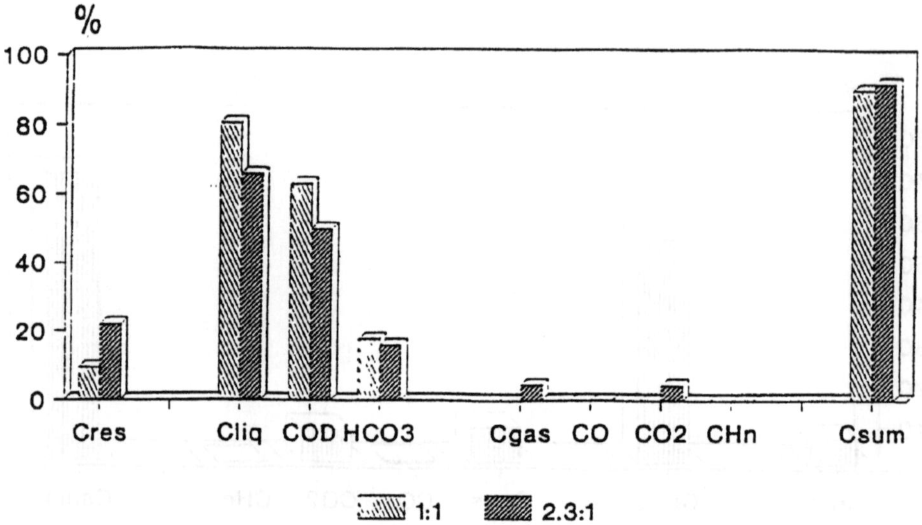

Fig. 7 C conversion ratio of P 5 after pressure hydrolysis with different alkaline conditions

4. Conclusion and Outlook

Under the conditions of alkaline pressure hydrolysis nitrocellulose is readily transformed to nitrate and nitrite and low chain carboxylic acids in the liquid phase. In the gaseous phase N_2, N_2O and NO_x are formed. Dependent on the alkaline content the NO_x amount can be reduced below 0.1 % relating to the nitrogen content of nitrocellulose. The results of the single base propellant P 1 show that the additive DPA and graphite remain as residue. Diphenyl-amine itself doesn't react under these conditions. Nevertheless it may be that some DPA may disappear as emulsion during the hydrolysis process. The same occures with centralite in the double-base propellants like P 2, 3, 4 and 5. The critical component in the double-base propellants is DNT. Here is not clear if all DNT is converted to harmless products. Additional investigations with pure 2,4-DNT show that only 7 % of the nitrogen was found as nitrite in the liquid phase. DNT on the other hand was not found in the solid residues. Therefore we assume that DNT is not completely converted and may be still present as an emulsion in the aqueous phase. The species DNT, DPA and centralite have to be removed from the reaction mixture before biological degradation will be performed. With acception of these species the alkaline pressure hydrolysis of propellants delivers an reaction mixture which is completely biodegradable. The two biological steps are annoxic dinitrification and normal aerobic treatment. The sludge after the biological steps is in the range of 5-8 % of raw material /4/.

treatment. The sludge after the biological steps is in the range of 5-8 % of raw material /4/.

For nitrocellulose based propellants alkaline pressure hydrolysis in combination with biological treatment seems to be an promising disposal process and an alternative to the common incineration.

For the critical components like centralite, DPA, DNT and TNT more drastically conditions have to be employed /5/. First experiments of supercritical hydrolysis and oxidation (above 374 °C and 221 bar) give good results /6/. It seems to be possible that under supercritical water conditions a complete destruction of the critical components could be achieved. In the case of hexachlorocyclohexane (HCH) containing pyrotechnic compositions the hydrolysis in presence of NaOH leads to conversions (Cl^-) higher 90 % after 30 minutes /7/.

5. References

/1/ Nitrocellulose, K. Fabel, 2. Bd., F. Enke Verlag, Stuttgart, S. 118 f.

/2/ W.O. Kenyon, H.L. Gray, J. Am. Chem. Soc. 1936 (58) 1422

/3/ Bunte, G.; Hirth, T.; Krause, H.
23. ICT Conference, June 30-July 3, 1992, Karlsruhe

/4/ Dahn, A.; Teller, M.
23. ICT Conference, June 30-July 3, 1992, Karlsruhe

/5/ Bunte, G.; Eisenreich, N.; Hirth, T.; Krause, H.
23. ICT Conference, June 30-Juli 3, 1992, Karlsruhe.

INDEX

MIX
Papier aus verantwortungsvollen Quellen
Paper from responsible sources
FSC® C105338

If you have any concerns about our products,
you can contact us on
ProductSafety@springernature.com

In case Publisher is established outside the EU,
the EU authorized representative is:
Springer Nature Customer Service Center GmbH
Europaplatz 3, 69115 Heidelberg, Germany

Printed by Libri Plureos GmbH
in Hamburg, Germany